ISBN 978-1-332-05171-7
PIBN 10276593

1 MONTH OF
FREE
READING

at
www.ForgottenBooks.com

By purchasing this book you are eligible for one month membership to ForgottenBooks.com, giving you unlimited access to our entire collection of over 700,000 titles via our web site and mobile apps.

To claim your free month visit:
www.forgottenbooks.com/free276593

CATALOGUE

OF THE

SPECIMENS

OF

HEMIPTERA HETEROPTERA

IN

THE COLLECTION

OF THE

BRITISH MUSEUM.

—

PART VI.

—

BY

FRANCIS WALKER.

PRINTED FOR THE TRUSTEES OF THE BRITISH MUSEUM:
LONDON, 1873.

LONDON:

E. NEWMAN, PRINTER, DEVONSHIRE STREET, BISHOPSGATE.

PREFACE.

The object of the present Catalogue is to give a complete List of all the genera and species of Hemiptera Heteroptera known to exist in the collections of European and American Entomologists. The letters *a*, *b*, *c*, &c., after the species, denote the specimens now contained in the British Museum, followed by the *habitat* and the mode in which each of them was obtained ; and the absence of these *letters* indicates the species which are desiderata, and therefore desirable to be procured for the collection.

J. E. GRAY.

British Museum,
    January 21st, 1873.

CATALOGUE

OF

HEMIPTERA HETEROPTERA.

———

PART VI.

———

Fam. PYRRHOCORIDÆ (continued).

Genus 7. DINDYMUS.

Dindymus, *Stal, Berl. Ent. Zeit.* vii. 397.

South Asia and Eastern Isles.

A. Eyes not petiolated.
a. Head with a more or less distinct neck.
  * Head very convex above and beneath.
    † Neck longer. -   -   -   -   -   sphærocephalus.
  †† Neck shorter.
    ‡ Body less elongated. -   -   -   -   fecialis.
    ‡‡ Body more elongated.   -   -   -   crudelis.
  ** Head slightly convex.
    † Head red.
    ‡ Prothorax red.
    § Sides of prothorax much reflexed.
    × Membrane black at the tip. -   -   -   sanguineus.
  ×× Membrane pale at the tip. -   -   -   rubiginosus,
    §§ Sides of the prothorax very slightly reflexed. -   discoidalis.
    ‡ Prothorax black.
    § First and second joints of the antennæ black. -   bicolor.
    §§ First and second joints of antennæ pale.   -   mundus.
  †† Head black.
    ‡ A pale band across the scutellum and the base
      of the corium.   -   -   -   -   basifer.
    ‡‡ No pale band across the scutellum.

§ Prothorax wholly black.
✕ Prothorax excavated on each side.
o Prothorax contracted in front.
⊷ Corium black at the base.   -   -   -   venustus.
⊹⊹ Corium not black at the base. -   -   -   decisus.
oo Prothorax not contracted in front.   -   -   clarus.
✕✕ Prothorax not excavated on each side.
  o Legs wholly black.   -   -   -   -   reduvioides.
  oo Knees and tarsi pale. -   -   -   -   intermedius.
§§ Prothorax not wholly black.
✕ Prothorax not white in front.
o Prothorax not brown.
⊷ Hind border of the prothorax whitish.   tricolor.
⊹⊹ Hind border of the prothorax luteous. -   -   venustulus.
⊹⊹⊹ Hind border of the prothorax not whitish nor
    luteous.   -   -   ··   -   -   varius.
oo Prothorax brown.   -   ·   -   -   effusus.
✕✕ Prothorax white in front.
  o Prothorax black.   -   ··   -   -   rutilans.
  oo Prothorax red.   -   ··   -   -   simplex.
b. Head rather abruptly contracted behind, not
    received by the prothorax as far as the eyes.
  * Body oblong.
  † Sides of the prothorax like the disk in colour.
  ‡ Antennæ whitish.   -   -   -   -   albicornis.
  ‡‡ Antennæ black.
  § Hind border of the prothorax not white.
  ✕ Prothorax with two spots.   -   -   -   pulcher.
✕✕ Prothorax with no spots.   -   -   -   semirufus.
  §§ Hind border of the prothorax white. -   -   Amboinensis.
  †† Sides of the prothorax not like the disk in
    colour.   -   -   -   -   -   vinulus.
  ** Body broader, oval.
    (" This subdivision approaches Cenæus,
    which genus differs from Dindymus in
    the structure of the antennæ, and per-
    haps hardly ought to be separated from
    it."—Stal.)
  † First and second joints of the antennæ black. ··   ovalis.
  †† First and second joints of the antennæ pale. -   lanius.

1. DINDYMUS LANIUS.

lanius, *Stal, Berl. Ent. Zeit.* vii. 401 ; *K. Sv. Vet. Ak. Handl.* 1870, 113.
Hindostan.

2. DINDYMUS OVALIS.

ovalis, *Stal, Berl. Ent. Zeit.* vii. 401 ; *K. Sv. Vet. Ak. Handl.* 1870, 112.
Hindostan.

3. DINDYMUS SANGUINEUS.

Cimex augur *var. β, Thunb. Nov. Sp. Ins.* iii. 58—**Lygæus** sanguineus, *Fabr. Ent. Syst.* iv. 155—Lygæus cruentus, *Fabr. Syst. Rhyn.* 223 —Lygæus hæmatideus, *Fabr. Syst. Rhyn.* 223—Pyrrhocoris hæmatideus, *Hahn, Wanz. Ins.* i. 9, pl. 1, f. 3—Dysdercus Augur, *Stal, Ofv. K. V. Ak. Forh.* 1855, 371—Dindymus Augur, *Stal, Berl. Ent. Zeit.* vii. 399—Dindymus hæmatideus, *Mayr, Reise, Nov. Hem.* 130 —Dindymus sanguineus, *Stal, Hem. Fabr.* i. 82; *K. Sv. Vet. Ak. Handl.* 1870, 112.

a—d. Hong Kong. Presented by J. C. Bowring, Esq.
e—i. China. Presented by G. T. Lay, Esq.
j, k. Hindostan. Presented by J. C. Bowring, Esq.

4. DINDYMUS AMBOINENSIS.

Lygæus Amboinensis, *Fabr. Syst. Rhyn.* 225—Dindymus Amboinensis, *Stal, Berl. Ent. Zeit.* vii. 400 ; *K. Sv. Vet. Ak. Handl.* 1870, 111— Dindymus tricolor, *Mayr. Verh. Zool. Bot. Ges. Wien.* xv. 436.

Amboina. Ternate.

a, b. China. Presented by G. T. Lay, Esq.
c. Ceram. Presented by W. W. Saunders, Esq.
d, e. Ceram. From Madame Ida Pfeiffer's collection.

5. DINDYMUS VINULUS.

Dindymus vinulus, *Stal, Berl. Ent. Zeit.* vii. 400; *Ofv. K. V. Ak. Forh.* 1870, 666 ; *K. Sv. Vet. Ak. Handl.* 1870, 110.

Philippine Isles.

6. DINDYMUS PYROCHROUS.

Dysdercus pyrochrous, *Boisd. Voy. Astr. Ent.* ii. 642, pl. 11, f. 9— Dindymus pyrochrous, *Stal, K. Sv. Vet. Ak. Handl.* 1870, 112.

Aru. Mysol.

7. DINDYMUS FECIALIS.

Dindymus fecialis, *Stal, Berl. Ent. Zeit.* vii. 397; *Ofv. K. V. Ak. Forh.* 1870, 666; *K. Sv. Vet. Ak. Handl.* 1870, 112.

a. Philippine Isles. From Dr. Cuming's collection.

8. DINDYMUS MUNDUS.

Dindymus mundus, *Stal. Berl. Ent. Zeit.* vii. 398; *Ofv. K. V. Ak. Forh.* 1870, 666; *K. Sv. Vet. Ak. Handl.* 1870, 110.

Philippine Isles.

9. DINDYMUS VENUSTUS.

Dindymus venustus, *Stal, Berl. Ent. Zeit.* vii. 398; *K. Sv. Vet. Ak. Handl.* 1870, 110; *Ofv. K. V. Ak. Forh.* 1870, 666.

a. Philippine Isles. From Dr. Cuming's collection.

10. DINDYMUS PULCHER.

Dindymus pulcher, *Stal, Berl. Ent. Zeit.* vii. 400; *Ofv. K. V. Ak. Forh.* 1870, 666; *K. Sv. Vet. Ak. Handl.* 1870, 110.

Philippine Isles.

11. DINDYMUS CRUDELIS.

crudelis, *Stal, Berl. Ent. Zeit.* vii. 397; *K. Sv. Vet. Ak. Handl.* 1870, 111.

Celebes.

a. ? Philippine Isles.   From Dr. Cuming's collection.

12. DINDYMUS RUBIGINOSUS.

Cimex rubiginosus, *Fabr. Mant. Ins.* ii. 301; *Gmel. ed. Syst. Nat.* i. 2174 —Lygæus rubiginosus, *Fabr. Ent. Syst.* iv. 159; *Syst. Rhyn.* 226— Pyrrhocoris rubiginosus, *Burm. Handb. Ent.* ii. 284—Dysdercus hypogastricus, *H.-Sch. Wanz. Ins.* ix. 177, pl. 317, f. 979—Dindymus rubiginosus, *Stal, Berl. Ent. Zeit.* vii. 399; *K. Sv. Vet. Ak. Handl.* 1870, 111.

a. Java.   From the East India Company's collection.
b, c. Sumatra.
d. Lombok.   Presented by W. W. Saunders, Esq.
e. Celebes.
f, g. Celebes.   From Mr. Wallace's collection.

13. DINDYMUS BICOLOR.

Cimex augur, var. δ, *Thunb. Nov. Sp. Ins.* iii. 58, f. 69—Pyrrhocoris bicolor, *H.-Sch. Wanz. Ins.* vi. 27, pl. 189, f. 585—Dysdercus thoracicus, *Stal, Ofv. K. V. Ak. Forh.* 1855, 391—Dindymus bicolor, *Stal, Berl. Ent. Zeit.* vii. 398; *K. Sv. Vet. Ak. Handl.* 1870, 111.

a. Java.   From the East India Company's collection.

14. DINDYMUS ALBICORNIS.

Lygæus albicornis, *Fabr. Syst. Rhyn.* 223—Dindymus albicornis, *Stal, Berl. Ent. Zeit.* vii. 399; *K. Sv. Vet. Ak. Handl.* 1870, 110.

China.

a. Java.   From the East India Company's collection.
b. Philippine Isles.   From Dr. Cuming's collection.

15. DINDYMUS SEMIRUFUS.

semirufus, *Stal, Berl. Ent. Zeit.* vii. 400; *K. Sv. Vet. Ak. Handl.* 1870, 111.

Cambodia.

16. DINDYMUS SPHÆROCEPHALUS.

Dindymus sphærocephalus, *Stal, Berl. Ent. Zeit.* vii. 397; *Ofv. K. V. Ak. Forh.* 1870, 665; *K. Sv. Vet. Ak. Handl.* 1870, 112.

a. Philippine Isles. From Dr. Cuming's collection.

17. DINDYMUS THUNBERGI.

Cimex augur, *var. γ, Thunb. Nov. Ins. Sp.* iii. 58, f. 68—Dysdercus Thunbergi, *Stal, Ofv. Vet. Ak. Forh.* 1855, 391—Dindymus Thunbergi, *Stal, K. Sv. Vet. Ak. Handl.* 1870, 111.

Hindostan.

18. DINDYMUS DECISUS.

Mas. *Niger, longi-ellipticus; rostrum segmentum 1um ventrale attingens; antennæ corpore breviores, articulo 4o albo apicem versus nigro; scutelli latera testacea; pectoris latera albo trifasciata; abdomen luteum; corium fulvum.*

Male. Black, elongate-elliptical. Head and under side shining. Rostrum extending to the first ventral segment. Antennæ somewhat shorter than the body; first joint much longer than the head; second a little shorter than the first; third much shorter than the second; fourth white for more than half the length from the base, a little longer than the second. Prothorax with a strongly-marked transverse middle furrow. Scutellum testaceous along each side. Pectus with three narrow white bands on each side. Abdomen luteous. Corium tawny. Length of the body 6 lines.

a. New Guinea. Presented by W. W. Saunders, Esq.

19. DINDYMUS BASIFER.

Mas. *Niger, fusiformis; rostrum coxas posticas attingens; antennæ corpore paullo breviores, articulo 4o ochraceo; scutellum et abdomen ochracea; segmenta pectoralia ochraceo marginata; corium ochraceum, nigro fasciatum, basi luteum; membrana fusca, basi cinerea.* Var. β. *—Segmenta pectoralia flavo marginata; abdomen luteum; corium apicem versus ferrugineum.*

Male. Black, fusiform, very finely punctured. Head and under side shining, smooth. Head elongate-triangular; middle lobe prominent. Eyes piceous, rather prominent. Rostrum extending to the hind coxæ. Antennæ a little shorter than the body; first joint very much longer than head; second very much shorter than the first; third a little shorter than the second; fourth ochraceous, a little longer than the second. Prothorax with a strongly-marked transverse furrow at a little in front of the middle; fore part smooth; sides reflexed. Scutellum, hind borders of the pectoral segments and abdomen ochraceous. Fore femora hardly incrassated, with three minute subapical spines beneath. Corium ochraceous, luteous at the base, and with a broad intermediate black band; membrane brown, cinereous at the base. *Var. β.*—Hind borders of the pectoral segments yellow. Abdomen luteous. Corium ferruginous, instead of ochraceous. Length of the body 8 lines.

a. Gilolo. Presented by W. W. Saunders, Esq.
b. Ternate. From Mr. Wallace's collection.

20. DINDYMUS DISCOIDALIS.

Fœm. *Rufus, fusiformis; rostrum coxas intermedias attingens, apice nigrum; antennæ corpore multo breviores, articulo 4o ochraceo; scutellum nigrum; mesopectus et metapectus nigro bimaculata; venter niger, rufo marginatus; membrana nigra, cinereo marginata.*

Female. Bright red, fusiform, shining. Head smooth, triangular; middle lobe prominent. Eyes lurid, rather prominent. Rostrum extending to the middle coxæ; tip black. Antennæ black, much shorter than the body; first joint longer than the head; second longer than the first; third a little shorter than the first; fourth ochraceous, a little longer than the second. Prothorax punctured, with two strongly-marked transverse furrows, of which the fore one is curved; space between the furrows smooth. Scutellum black. Mesopectus and metapectus with a large black spot on each side. Abdomen beneath black, except on each side and towards the tip. Fore femora slightly incrassated, with a few minute spines beneath. Corium punctured; membrane black, with a narrow cinereous border. Length of the body 8 lines.

a. Philippine Isle. From Dr. Cuming's collection.

21. DINDYMUS CLARUS.

Fœm. *Niger, fusiformis; rostrum segmentum 1um ventrale attingens; antennæ corpore multo breviores, articulo 4o luteo apice nigro; metapectus luteo marginatum; venter luteus; corium rufum, basi nigrum; membrana basi cinerea.*

Female. Black, fusiform, shining. Rostrum extending to the first ventral segment. Antennæ much shorter than the body; first joint very much longer than the head; second very much shorter than the first; third rather shorter than the second; fourth luteous, black at the tip, hardly shorter than the second. Prothorax with a strongly-marked transverse middle furrow at a little in front of the middle; sides much reflexed. Metapectus with a luteous border. Abdomen luteous beneath. Corium red-lead colour, black at the base. Membrane cinereous at the base. Hind wings brown. Length of the body 7½ lines.

a. Celebes. Presented by W. W. Saunders, Esq.

22. DINDYMUS SIMPLEX.

Mas. *Niger, fere linearis; rostrum segmenti 1i ventralis marginem posticum attingens; antennæ corpore breviores; prothorax, pectus, abdomen et corium rufa; prothoracis margo anticus albus; pectus nigro fasciatum, segmentorum marginibus posticis albis; segmenta ventralia 2um et 3um albido marginata; corium linea apud marginem interiorem nigra; membrana cinereo marginata.*

Male. Black, nearly linear. Rostrum extending to the hind border of the first ventral segment. Antennæ rather shorter than the body ; first joint much longer than the head ; second very much shorter than the first ; third very much shorter than the second ; fourth as long as the first. Prothorax, pectus, abdomen and corium bright red. Prothorax with a strongly-marked transverse furrow at one-third of the length from the fore border, which is white. Pectus with a black band ; hind borders of the segments white. Hind borders of the second and third ventral segments whitish. Fore femora beneath with two subapical spines. Corium with a black line along the interior border. Membrane at the tip with a narrow pale cinereous border. Length of the body 6 lines.

a. Celebes. Presented by W. W. Saunders, Esq.

23. DINDYMUS IMITATOR.

Mas. *Rufus, fusiformis, capite subtus rostro antennis, pectore pedibus membranaque nigris ; rostrum segmentum 1um ventrale attingens ; antennæ corpore multo breviores ; membrana basi albida.*

Male. Red, fusiform, very finely punctured. Head beneath, rostrum, antennæ, pectus, legs and membrane black. Head triangular. Eyes piceous, hardly petiolated. Rostrum extending to half the length of the first ventral segment. Antennæ much shorter than the body ; first joint longer than the head, red towards the base ; second much shorter than the first ; third much shorter than the second. Prothorax with a strongly-marked transverse callus at somewhat in front of the middle, and with an anterior transverse callus ; sides slightly reflexed. Membrane whitish at the base. Length of the body 5 lines.

a. Siam. Presented by W. W. Saunders, Esq.

24. DINDYMUS EFFUSUS.

Mas. *Piceus, fusiformis, tomentosus, capite antennis pectoreque nigris ; rostrum testaceum, segmenti 1i ventralis marginem posticum attingens ; antennæ corpore multo breviores, articulo 4o albo apice nigro ; prothorax lateribus margineque postico pallide flavis ; venter rufus ; femora basi flava ; corium ferrugineum, plaga subapicali flavescente ; membrana diaphana.*

Male. Piceous, fusiform, tomentose, very finely punctured. Head, antennæ and pectus black. Head elongate-triangular ; middle lobe prominent. Eyes prominent. Rostrum testaceous, extending to the hind border of the first ventral segment. Antennæ much shorter than the body ; first joint much shorter than the head ; second shorter than the first ; third much shorter than the second ; fourth white, black at the tip, nearly as long as the second. Prothorax bordered with pale yellow on each side and behind ; two distinct transverse furrows, one at somewhat in front of the middle, the other near the fore border. Abdomen red beneath. Femora towards the base and trochanters yellow ; fore femora hardly incrassated. Corium of the fore wings ferruginous, with a yellowish patch near the tip, which is sometimes brown; membrane pellucid. Length of the body 7 lines.

a. Mysol. From Mr. Wallace's collection.
b, c. Dory, New Guinea. Presented by W. W. Saunders, Esq.
d—f. Dory, New Guinea. *From Mr. Wallace's collection.*

25. DINDYMUS VENUSTULUS.

Mas. *Niger, fusiformis; rostrum segmenti ventralis 1i marginem posticum attingens; antennæ corpore sat breviores, articuli 4i dimidio basali albo; prothorax antice albo postice luteo marginatus; venter luteo marginatus et univittatus; pedes femoribus anticis subincrassatis, femoribus tarsisque quatuor posterioribus basi albis; corii costa fasciaque subapicalis albæ.* Var. β.—*Prothorax albo marginatus; antennarum articulus 4us totus niger.*

Male. Black, fusiform, hardly punctured. Head triangular, shining, smooth. Eyes very prominent. Rostrum extending to the hind border of the first ventral segment. Antennæ somewhat shorter than the body; first joint as long as the head; second a little shorter than the first; third much shorter than the second; fourth a little shorter than the first, white for nearly half the length from the base. Prothorax with a well-defined transverse middle furrow, and with a narrow border which is luteous behind and white in front. Abdomen beneath with a luteous border and a broad irregular luteous stripe. Fore femora slightly incrassated, with subapical spines beneath; four posterior femora and tarsi white towards the base. Corium with a white costa and a white subapical band. *Var. β.*—Prothorax with a white border. Fourth joint of the antennæ wholly black; stripe beneath the abdomen indistinct. Length of the body 4—4½ lines.

a. New Guinea. Presented by W. W. Saunders, Esq.
b. New Guinea. From Mr. Wallace's collection.

26. DINDYMUS RUTILANS.

Mas. *Niger, fusiformis; rostrum segmentum 2um ventrale attingens; antennæ corpore multo breviores; prothorax postice fortiter punctatus, margine antico albo; segmenta pectoralia albo marginata; abdomen coriique dimidium basale rufa; membrana diaphana.*

Male. Black, fusiform, smooth, shining. Head triangular. Eyes very prominent. Rostrum extending to the second ventral segment. Antennæ much shorter than the body; first joint longer than the head; second much shorter than the first; third shorter than the second. Prothorax with a strongly-marked transverse furrow at somewhat in front of the middle; hind part roughly punctured; fore border white. Pectoral segments bordered with white. Abdomen bright red. Corium bright red for half the length from the tip. Membrane pellucid. Length of the body 4½ lines.

a. Siam. Presented by W. W. Saunders, Esq.

27. DINDYMUS REDUVIOIDES.

Mas et fœm. Niger, fusiformis; rostrum coxas posticas attingens; antennæ corpore multo breviores, articulo 4o luteo apice nigro; metapectus albo marginatum; abdominis dorsum rufum; venter rufo marginatus; corium rufum, basi nigrum; membrana basi cinerea.

Male and female. Black, fusiform, smooth, shining. Head elongate. Eyes piceous, not prominent, remote from the prothorax. Rostrum extending to the hind coxæ. Antennæ much shorter than the body; first joint much longer than the head; second very much shorter than the first; third shorter than the second; fourth luteous, black at the tip, longer than the second. Prothorax with a strongly-marked transverse furrow at somewhat before the middle; hind part thinly punctured; sides reflexed. Metapectus bordered with white. Abdomen red above and along each side and at the tip beneath. Legs slender; fore femora beneath with two very minute subapical spines. Corium of the fore wings bright red, black towards the base; membrane cinereous at the base. Length of the body 7½—10 lines.

a—c. Celebes. Presented by W. W. Saunders, Esq.
d. Celebes. From Madame Ida Pfeiffer's collection.
e. Celebes. . From Mr. Wallace's collection.

28. DINDYMUS INTERMEDIUS.

Mas. Niger, fusiformis; caput cyaneo-nigrum; rostrum apicem versus rufum, coxas posticas attingens; antennæ corpore breviores, articulo 4o luteo apicem versus nigro; pectus et abdomen lutea, hujus disco postico nigro; pedes lutei, femoribus tibiisque posticis nigris; corium luteum.

Male. Black, fusiform. Head triangular, bluish black, smooth, shining. Eyes piceous, slightly prominent. Rostrum red towards the tip, extending to the hind coxæ. Antennæ rather shorter than the body; first joint much longer than the head; second much shorter than the first; third rather shorter than the second; fourth luteous, black towards the tip, nearly as long as the first. Prothorax velvety, with a slight transverse furrow near the hind border. Pectus and abdomen luteous; hind part of the disk of the latter black. Legs luteous; hind femora and hind tibiæ black, the former luteous towards the base. Fore wings abbreviated; corium luteous. Length of the body 5½ lines.

a. Mysol. From Mr. Wallace's collection.

29. DINDYMUS VARIUS.

Mas et fœm. Læte rufus, capite rostro antennis prothorace antice pedibusque nigris; rostrum segmenti 1i ventralis marginem posticum attingens; antennæ corpore breviores, articulo 4o albo apicem versus nigro; scutellum basi nigrum; pectus album, nigro trifasciatum; venter pallide testaceus; membrana lurida. Var. β.—Membrana apicem versus nigra. Var. γ.—Membrana nigra. Var. δ.—Scutellum nigrum. Var. ε.—Corium apud marginem interiorem nigrum.

Male and female. Bright red. Head, rostrum, antennæ, fore division of the prothorax and legs black. Head shining. Eyes piceous, hardly prominent, at some distance from the prothorax. Rostrum extending to the hind border of the first ventral segment. Antennæ shorter than the body; first joint very much longer than the head; second much shorter than the first; third a little shorter than the second; fourth white, black towards the tip, longer than the third. Prothorax black in front of the strongly-marked transverse furrow; sides reflexed. Scutellum black at the base. Pectus mostly white, with three black bands. Abdomen beneath pale testaceous. Fore femora beneath with three subapical spines. Membrane of the fore wings lurid. *Var. β.*—Membrane black towards the tip. *Var. γ.*—Membrane wholly black. *Var. δ.*—Scutellum wholly black. *Var. ε.*—Corium with an elongated black patch along the interior border. Length of the body 6—9 lines.

a—c. Aru. Presented by W. W. Saunders, Esq.
d—g. Aru. From Mr. Wallace's collection.
h, i. Ké. Presented by W. W. Saunders, Esq.
j. Ké. From Mr. Wallace's collection.
k. New Guinea. Presented by W. W. Saunders, Esq.
l, m. New Guinea. From Mr. Wallace's collection.

30. DINDYMUS INDIGNUS.

Mas. *Niger, fusiformis; capite supra connexivo ventreque fulvis; rostrum coxas posticas paullo superans; prothorax fulvo tenuiter marginatus.*

Male. Black, fusiform, very finely punctured. Head triangular, tawny above; middle lobe very prominent. Eyes very prominent. Rostrum extending to a little beyond the hind coxæ. Prothorax with a strongly-marked transverse furrow at one-third of the length from the fore border, and with a narrow tawny border. Abdomen tawny beneath, and on each side above. Legs rather stout; fore femora beneath with minute subapical spines. Length of the body 7 lines.

a. Siam. Presented by W. W. Saunders, Esq.

31. DINDYMUS VENTRALIS.

ventralis, *Mayr, Verh. Zool. Bot. Gesell. Wien.* xvi. 364; *Reise, Nov. Hem.* 132, pl. 3, f. 33; *Stal, K. Sv. Vet. Ak. Handl.* 1870, 113.

Philippine Isles. Australia.

Australia.

A. Corium with no black points.
  a. Abdomen with no black bands beneath.
    * Hind part of the prothorax luteous. -   -   -   versicolor.
  ** Hind part of the prothorax black. -   -   - circumcinctus.
    b. Abdomen with black bands beneath. -   -   -   cinctifer.
B. A black point on the corium. -   -   -   - bipunctatus.

32. DINDYMUS CIRCUMCINCTUS.

circumcinctus, *Stal, Berl. Ent. Zeit.* vii. 400; *K. Sv. Vet. Ak. Handl.* 1870, 113; *Mayr, Reise, Nov. Hem.* 131, pl. 3, f. 31, 32.

Moreton Bay.

Var.? Mas. *Niger, fusiformis; rostrum coxas posticas attingens; antennæ corpore paullo breviores, articulo 4o basi albo; prothoracis latera rufa; segmenta pectoralia albo marginata; connexivum rufum; tibiæ tarsique saturate rufa; corium costa margineque interiore rufis.*

Male. Black, fusiform, shining, very finely punctured. Head elongate-triangular. Eyes piceous, rather prominent. Prothorax extending to the hind coxæ. Antennæ a little shorter than the body; first joint much longer than the head; second a little shorter than the first; third a little shorter than the second; fourth white at the base, a little longer than the second. Prothorax with a strongly-marked transverse middle furrow; sides red. Pectoral segments bordered with white. Connexivum red. Fore femora beneath with two minute subapical spines. Tibiæ and tarsi dark red. Corium red along the costa, and more broadly red along the interior border. Length of the body 4 lines.

a. Australia. From Mr. Damel's collection.

33. DINDYMUS VERSICOLOR.

Odontopus versicolor, *H.-Sch. Wanz. Ins.* ix. pl. 315, f. 969—Dindymus versicolor, *Stal, K. Sv. Vet. Ak. Handl.* 1870, 113.

a—g. Tasmania. Presented by W. W. Saunders, Esq.
h. New South Wales. Presented by W. W. Saunders, Esq.
i. Australia. From Mr. Damel's collection.
j. South Australia.
k. South Australia. Presented by G. F. Angas, Esq.
l—o. Australia. From the late Earl of Derby in 1851.
p—s. Tasmania. From Dr. Hooker's collection.
t—w. New South Wales. Presented by the Earl of Derby in 1845.
x, y. Adelaide. Presented by the Entomological Club.
z. South Australia. Presented by R. Bakewell, Esq.

34. DINDYMUS BIGUTTATUS.

Mas. *Ochraceo-rufus; rostrum segmenti 1i ventralis marginem posticum attingens; antennæ nigræ, nonnunquam ex parte aut omnino rufæ; scutelli discus nonnunquam niger; pectus nonnunquam nigro quadriguttatum, segmentis nonnunquam testaceo marginatis; venter testaceus, segmentis rufo marginatis; tibiæ tarsique nigra; corium nigro uniguttatum; membrana nigra, nonnunquam ex parte aut omnino albida.*

Male. Orange-red, fusiform. Rostrum extending to the hind border of the first ventral segment; tip black. Antennæ black, occasionally partly or wholly red; first joint much longer than the head; second rather shorter than the first; third much shorter than the second. Prothorax with a well-defined transverse furrow at one-third of the length from the fore border. Disk of the scutellum occasionally black. Pectus occasionally with two black dots on each side; hind borders of the segments sometimes pale testaceous. Abdomen beneath pale testaceous; fore borders of the segments red. Tibiæ and tarsi black. Corium with a black dot in the disk at a little beyond the middle. Membrane black, sometimes partly or wholly whitish. Length of the body 4½—7½ lines.

a, b. Port Essington. Presented by the Earl of Derby in 1846.
c—h. Victoria River, North Australia. Presented by J. R. Elsey, Esq.

35. DINDYMUS CINCTIFER.

Mas. *Læte rufus, fusiformis, tomentosus, rostro antennis scutello pectore pedibus membranaque nigris; rostrum segmenti 1i ventralis marginem posticum attingens; antennæ corpore multo breviores; scutellum apice rufum; segmenta ventralia nigro late marginata; membrana cinereo pallido late marginata.*

Male. Bright red, fusiform, tomentose. Head shining, triangular; middle lobe slightly prominent. Eyes piceous, slightly prominent. Rostrum, antennæ, pectus and legs black. Rostrum extending to the hind border of the first ventral segment. Antennæ much shorter than the body; first joint much shorter than the head; second about four times the length of the first; third longer than the second. Prothorax with a well-defined transverse furrow near the fore border, and with a slight transverse furrow on each side very near the hind border. Scutellum black, red towards the tip. Abdomen beneath with a broad black band near the hind border of each segment. Fore femora beneath with two minute subapical spines. Membrane black, broadly bordered with pale cinereous. Length of the body 7½ lines.

a. Moreton Bay. From Mr. Diggles' collection.

Country unknown.
36. DINDYMUS DUBIUS.

Mas. *Niger, fusiformis; rostrum coxas posticas attingens; antennæ subclavatæ, corpore multo breviores; prothoracis latera lutea; segmenta pectoralia albo marginata; connexivum rufum; tibiæ saturate rufæ; corium costa margineque interiore basi fulvis.*

Male. Black, fusiform. Head and under side shining. Rostrum extending to the hind coxæ. Antennæ much shorter than the body; first joint luteous towards the base, much longer than the head; second much shorter than the first; third subclavate, a little shorter than the second. Prothorax with a strongly-marked transverse furrow; sides luteous, reflexed. Pectoral segments with white hind borders. Connexivum red. Tibiæ dark red. Costa and interior border of the corium tawny towards the base. Membrane not developed. Hind wings cinereous. Length of the body 5 lines.

a. ——— ?

Genus 8. MELAMPHAUS.

Melamphaus, *Stal, Hem. Fabr.* i. 83.

A. Rostrum extending to the middle coxæ.
a. Prothorax black.
* Prothorax ochraceous along the fore border. - - circumdatus.
** Prothorax not ochraceous along the fore border. - lyroides.
b. Prothorax red.
* Corium black. - - - - - angulifer.
** Corium red. - - - - - - scutifes.
B. Rostrum extending to the ventral segments.
a. Corium black, with a red costa.
* Prothorax bordered with red. - - - lateralis.
** Prothorax not bordered with red. - - - costalis.
b. Corium black, with a luteous costa. - - - marginalis.
c. Corium black, with a red spot or patch. - - faber.
d. Corium red. - - - - - - rubidus.
e. Corium luteous, partly black. - - - femoratus.

1. MELAMPHAUS FABER.

Cimex faber, *Fabr. Mant. Ins.* ii. 297; *Gmel. ed. Syst. Nat.* i. 2188—
Lygæus faber, *Fabr. Ent. Syst.* iv. 146; *Syst. Rhyn.* 215—Melam-
phaus faber, *Stal, Hem. Fabr.* i. 83; *Ofv. K. V. Ak. Forh.* 1870, 667;
K. Sv. Vet. Ak. Handl. 1870, 108.

This species is very variable as to markings.

a. Silbet. From the Rev. Mr. Stainsforth's collection.
a, b. Hindostan. Presented by Major-Gen. Hardwicke.
c—f. Philippine Isles. From Dr. Cuming's collection.

2. MELAMPHAUS FULVOMARGINATUS.

Dysdercus fulvomarginatus, *Dohrn, Stett. Ent. Zeit.* xxi. 405.—Melamphaus fulvomarginatus, *Stal, K. Sv. Vet. Ak. Handl.* 1870, 108.

Ceylon.

3. MELAMPHAUS RUBROCINCTUS.

Dysdercus? rubrocinctus, *Stal, Berl. Ent. Zeit.* vii. 403—Melamphaus
rubrocinctus, *Stal, K. Sv. Vet. Ak. Handl.* 1870, 108.

Assam.

4. MELAMPHAUS LATERALIS.

Mas. *Niger, fusiformis, cinereo tomentosus ; rostrum segmentum 2um
ventrale attingens ; antennæ corpore breviores, articulo 4o basi luteo ;
prothorax et segmenta pectoralia rufo marginata ; abdomen rufum,
segmentorum ventralium marginibus anticis nigris ; corii costa rufa.*

Male. Black, fusiform, with cinereous tomentum. Head triangular.
Eyes slightly prominent. Rostrum extending to the middle of the second
ventral segment. Antennæ slender, shorter than the body; first joint

very much longer than the head; second shorter than the first; third shorter than the second; fourth a little longer than the third, luteous towards the base. Prothorax with a narrow red margin, which is more distinct on each side than on the fore border and on the hind border; a well-defined transverse furrow at somewhat before the middle. Pectoral segments bordered with red. Abdomen red; fore borders of the ventral segments black. Legs long, slender; fore femora beneath with three subapical spines. Corium of the fore wings with a red costa. Length of the body 8—10 lines.

a, b. Ceylon. Presented by Dr. Templeton.
c. Ceylon. From Dr. Thwaites' collection.

5. MELAMPHAUS COSTALIS.

Mas. *Niger, fusiformis; rostrum segmenti 1i ventralis marginem posticum attingens; antennæ corpore multo breviores, articulo 4o basi albo; abdomen et corii costa ochracea.*

Male. Black, fusiform. Rostrum extending to the hind border of the first ventral segment. Antennæ much shorter than the body; first joint much longer than the head; second a little shorter than the first; third much shorter than the second; fourth white towards the base, as long as the second. Prothorax punctured, with a well-defined transverse furrow at one-third of the length from the fore border. Abdomen and costa of the corium red-lead colour. Length of the body 7 lines.

a. Celebes. Presented by W. W. Saunders, Esq.

6. MELAMPHAUS MARGINALIS.

Mas. *Niger, fusiformis; rostrum segmenti 2i ventralis marginem posticum attingens; antennæ corpore breviores; prothorax lateribus margineque postico luteis, illis vix reflexis; segmenta pectoralia rufo marginata; venter rufus; corii costa lutea.*

Male. Black, fusiform. Head elongate-triangular. Eyes hardly prominent, contiguous to the prothorax. Rostrum extending to the hind border of the second ventral segment. Antennæ slender, shorter than the body; first joint very much longer than the head; second much shorter than the first; third a little shorter than the second; fourth a little longer than the second. Prothorax with a well-defined transverse furrow at somewhat before the middle; sides and hind border luteous, the former very slightly elevated. Pectoral segments bordered with red. Abdomen red beneath. Legs long, slender; fore femora beneath with three subapical spines. Corium with a luteous costa. Length of the body 10—11 lines.

a, b. Ceylon. Presented by Dr. Templeton.

7. MELAMPHAUS RUBIDUS.

Mas. *Niger, fusiformis, capite prothoracisque lateribus saturate rufis, connexivo corioque rufis; rostrum segmenti 2i ventralis marginem posticum attingens; antennæ corporis dimidio longiores.*

Male. Black, fusiform, finely punctured. Head dark red, triangular. Eyes black, rather prominent. Rostrum extending to the hind border of the second ventral segment. Antennæ rather more than half the length of the body; first joint shorter than the head; second about twice the length of the first; third much shorter than the second; fourth a little shorter than the third. Prothorax with a slight transverse furrow at somewhat in front of the middle; sides dark red. Connexivum and corium red. Length of the body 7 lines.

a. Ceylon. Presented by Dr. Templeton.

8. MELAMPHAUS FEMORATUS.

Fœm. Luteus, fusiformis; capitis margine postico rostro *antennis scutello pedibus membranaque nigris; rostrum basi rufum, segmenti* 1i *ventralis marginem posticum fere attingens; antennæ corpore breviores; prothoracis discus niger; segmenta pectoralia albo marginata; abdominis segmenta nigro fasciata; femora rufa; corium striga apud marginem interiorem latissima maculaque triangulari discali maxima nigris.*

Female. Luteous, fusiform. Head triangular, black along the hind border. Eyes black, slightly prominent. Rostrum black, red at the base, extending nearly to the hind border of the first ventral segment. Antennæ black, slender, shorter than the body; first joint much longer than the head, red at the base; second shorter than the first; third shorter than the second; fourth as long as the third. Prothorax with a black disk; transverse furrow well defined; sides elevated. Scutellum black. Pectoral segments bordered with white. Abdomen with a black band on the fore border of each segment. Legs black, long, slender; coxæ, trochanters and femora bright red; fore femora beneath with three subapical spines. Corium with a very broad black streak along the inner border, and with a very large triangular black spot in the disk beyond; membrane black. Length of the body 10 lines.

a. North Hindostan. From Capt. Boyes' collection.

9. MELAMPHAUS ANGULIFER.

Fœm. Rufus, fusiformis, antennis tibiis tarsisque nigris; rostrum coxas intermedias attingens; antennæ corpore multo breviores; scutellum nigrum, apice rufum; metapectus nigro bimaculatum; venter vittis duabus nigris latis abbreviatis; corium nigrum, albido tenuiter marginatum.

Female. Bright red, fusiform, finely punctured. Eyes, antennæ and legs black. Eyes rather prominent. Rostrum black towards the tip, extending to the middle coxæ. Antennæ much shorter than the body; first joint longer than the head; second a little longer than the first; third much shorter than the first; fourth much longer than the second. Prothorax with two well-defined transverse furrows; fore furrow much curved; sides slightly reflexed. Scutellum black, red at the tip. Metapectus with a large black spot on each side. Abdomen beneath with a broad and much abbreviated black stripe on each side. Femora red. Fore femora slightly incrassated, with two minute spines near the tips beneath. Corium black, with a very narrow whitish border. Length of the body 8 lines.

a. Ceram.   Presented by W. W. Saunders, Esq.
b. Ceram.   From Madame Ida Pfeiffer's collection.

10. MELAMPHAUS SCUTIFER.

Fœm. Læte rufus, fusiformis, rostro antennis pedibusque nigris; rostrum coxas intermedias attingens; antennæ corpore breviores; prothorax nigro unifasciatus; scutellum nigrum; pectus nigro sex plagiatum; venter niger rufo marginatus; membrana nigra, cinereo marginata.

Female. Bright red, fusiform. Head triangular. Eyes lurid, prominent, very near the prothorax. Rostrum, antennæ and legs black. Rostrum extending to the middle coxæ. Antennæ somewhat shorter than the body; first joint much longer than the head; second longer than the first; third shorter than the first; fourth as long as the second. Prothorax with two distinct transverse furrows, one in the middle, the other near the fore border and much curved; space between them black; sides reflexed. Scutellum black. Pectus with three large black patches on each side. Abdomen beneath black, red along each side and towards the tip. Legs rather stout; fore femora slightly incrassated, with minute subapical spines beneath. Membrane of the fore wings black, with a pale cinereous border. Length of the body 9 lines.

a. Gilolo.   Presented by W. W. Saunders, Esq.

11. MELAMPHAUS CIRCUMDATUS.

Fœm. Niger, fusiformis, subtus rufus; rostrum coxas intermedias attingens; antennæ corpore breviores; prothorax ochraceo late marginatus; pedes longi, graciles, coxis et trochanteribus ochraceis; corii costa ochraceo vittata.

Female. Black, fusiform, red beneath. Head shining, triangular. Eyes slightly prominent, contiguous to the prothorax. Rostrum extending to the middle coxæ. Antennæ slender, shorter than the body; first joint nearly twice the length of the head; second a little shorter than the first; third shorter than the second; fourth as long as the first. Prothorax broadly ochraceous along each side, more broadly so along the fore border; two transverse furrows, one at a little in front of the middle more strongly marked than the other, which is angular and near the fore border. Mesopectus and metapectus mostly black and shining. Legs black, long, slender; coxæ and trochanters ochraceous; fore femora beneath with two subapical spines. Corium with an ochraceous costal stripe. Length of the body 8—8½ lines.

a. Wagiou.   Presented by W. W. Saunders, Esq.
b. Aru.   Presented by W. W. Saunders, Esq.
c, d. Aru.   From Mr. Wallace's collection.

12. Melamphaus lycoides.

Fœm. *Niger, fusiformis, cinereo tomentosus; rostrum coxas intermedias attingens; antennæ corporis dimidio longiores; prothoracis latera rufa; scutellum apice rufescens; connexivum rufum; corii costa lutea; membrana nigro-fusco, apice cinerea.*

Female. Black, fusiform, with cinereous tomentum. Head triangular. Eyes rather prominent, very near the prothorax. Rostrum extending to the middle coxæ. Antennæ rather more than half the length of the body; first joint much longer than the head; second rather shorter than the first; third much shorter than the second; fourth longer than the third, shorter than the second. Prothorax with a slight transverse furrow at a little in front of the middle; sides red, reflexed. Scutellum reddish at the tip. Connexivum red. Fore femora slightly excavated, with four subapical spines beneath. Corium with a luteous costa. Membrane dark brown, cinereous at the tip. Length of the body 8 lines.

a. Philippine Isles. From Dr. Cuming's collection.

Div. 2.

Stictaulax, *Stal, K. Sv. Vet. Ak. Handl.* iv. 70, 107.

13. Melamphaus circumseptus.

Stictaulax circumsepta, *Stal, K. Sv. Vet. Ak. Handl.* 1870, 107.

Mysol. New Guinea.

Genus 9. PHYSOPELTA.

Physopelta, *Serv. Hist. Hem.* 271. *Stal, Hem. Afr.* iii. 2.

Africa.

A. Corium with a black spot. - - - - analis.
B. Corium with a black band. - - - festiva.

1. Physopelta analis.

Pyrrhocoris analis, *Sgnt. Arch. Ent.* ii. 306—Odontopus analis, *Stal, Ofv. K. V. Ak. Forh.* 1858, 441. Physopelta analis, *Stal, Hem. Afr.* iii. 2.

Calabar.

a. West Africa. From Mr. Stevens' collection.

2. Physopelta festiva.

Lygæus festivus, *Fabr. Syst. Rhyn.* 220—Physopelta festiva, *Stal, Hem. Fabr.* i. 79.

Guinea.

Var. Female. Black, fusiform, punctured. Head and prothorax shining. Head triangular. Eyes piceous, slightly prominent, near the prothorax.

Rostrum extending to the middle coxæ.   Antennæ much shorter than the
body ; first joint full twice the length of the head.   Prothorax slightly
convex and wholly red in front of the slight transverse furrow ; hind part
bordered with red.   Abdomen beneath testaceous, red in the disk, black at
the tip, and with three black streaks on each side.   Fore femora slightly
incrassated, beset beneath with a row of minute spines.   Corium of the
fore wings red, with a broad black stripe along the inner border, with a
broad black middle band, and with a black tip.   Length of the body
8½ lines.

a. Old Calabar.   From Mr. Gray's collection.

South Asia.

A. First joint of the antennæ longer than the second.
a. Corium with a white band.
  * Femora black beneath.   -   -   -   -   albifascia.
** Femora pale beneath.   -   -   -   -   robusta.
b. Corium with no white band. -   -   -   -   apicalis.
B. First and second joints of the antennæ equally long,
    or the first slightly shorter than the second.
a. Prothorax red.   -   -   -   -   -   Schlanbuschii.
b. Prothorax testaceous.   -   -   -   -   immunis.
c. Prothorax black.
  * Corium red or testaceous, with black spot.
  † First and second joints of the antennæ about equal in
    length.   -   -   -   -   -   - famelica, gutta.
†† First joint of the antennæ slightly shorter than the
    second.   -   -   -   -   -   - cincticollis.
** Corium black, red-bordered.   -   -   -   fimbriata.
C. Second joint of the antennæ longer than the first.   -   bimaculata.

3. PHYSOPELTA ALBOFASCIATA.

Cimex albofasciatus, *Deg. Ins.* iii. 335, pl. 34, f. 11—*Fabr. Sp. Ins.* ii. 364 ;
  Mant. Ins. ii. 299—Lygæus albofasciatus, *Fabr. Ent. Syst.* iv. 153 ;
  Syst. Rhyn. 221—Physopelta erythrocephala, *Serv. Hist. Hem.* 271—
  Physopelta affinis, *Serv. Hist. Hem.* 272—Physopelta albofasciata,
  Stal, Berl. Ent. Zeit. vii. 390.

a, b. Java.   From the East India Company's collection.
c. Java.   From Mr. Argent's collection.
d. Mount Ophir.   Presented by W. W. Saunders, Esq.

4. PHYSOPELTA ROBUSTA.

robusta, *Stal, Berl. Ent. Zeit.* vii. 390.

Laos.

a. Siam.   Presented by W. W. Saunders, Esq.

5. PHYSOPELTA GUTTA.

Pyrrhocoris gutta, *Burm. Nov. Act. Acad. Leop.* xvi. *Supp.* 300, pl. 41,
f. 10; *Handb. Ent.* ii. 285—Physopelta bimaculata, *Stal, Ofv. K. V.
Ak. Forh.* 1855, 186.

a. Silhet.   Presented by J. C. Bowring, Esq.
b. Pulo Penang.   Presented by J. C. Bowring, Esq.
c. Yang-Tsze.   Presented by J. C. Bowring, Esq.
d. Hong Kong.
e, f. Shanghai.   From Mr. Fortune's collection.
g. Ceylon.   Presented by Major Parry.
h. Silhet.   From Mr. Argent's collection.
i. Java.   From the East India Company's collection.
j. Philippine Isles.   From Dr. Cuming's collection.

6. PHYSOPELTA FAMELICA.

famelica, *Stal, Berl. Ent. Zeit.* vii. 391.

a. Ceram.   Presented by W. W. Saunders, Esq.
b. Gilolo.   Presented by W. W. Saunders, Esq.
c. Aru.   Presented by W. W. Saunders, Esq.
d. Aru.   From Mr. Wallace's collection.
e. Celebes.   Presented by W. W. Saunders, Esq.

Var.? *Nigra, fusiformis; caput rufum, margine postico nigro; antennæ
corpore multo breviores, basi rufæ; prothorax rufo marginatus;
venter rufus, guttis lateralibus nigris elongatis; femora antica rufa,
apice nigra; corium rufum, plaga postmedia maculaque apicali
nigris; membrana apice albo tenuiter marginata.*

Female. Black, fusiform. Head red; hind border above black.
Rostrum extending to the hind *coxæ.* Antennæ much shorter than the
body; first joint longer than the head, red at the base; second as long as
the first. Prothorax bordered with red in front on each side and more
narrowly behind; a distinct transverse middle furrow. Abdomen beneath
red, with a row of black elongated dots along each side. Fore femora red;
tips black. Corium red, with a black transverse patch beyond the middle
and with a black apical spot. Membrane with a narrow white apical border.
Length of the body 7½ lines.

a. Aru.   From Mr. Wallace's collection.

7. PHYSOPELTA CINCTICOLLIS.

cincticollis, *Stal, Berl. Ent. Zeit.* vii. 392.
Hindostan.

Var.? *Nigra, fusiformis; antennæ corpore multo breviores; prothorax
fascia postica nigra abbreviata latissima; venter rufus, apice niger;
coxæ et trochanteres lutea; corium rufum, nigro bifasciatum;
membrana apice cinereo marginata.*

Male. Black, fusiform. Rostrum extending to the hind coxæ. Antennæ much shorter than the body; first joint much longer than the head; second as long as the first; third much shorter than the second. Prothorax with a slight transverse middle furrow, and with an abbreviated and very broad black band on the hind part. Abdomen beneath red, black at the tip. Coxæ and trochanters luteous. Fore femora with four spines beneath. Corium red, with two irregular black bands, one a little beyond the middle, the other subapical. Membrane with a pale cinereous apical border. Length of the body 7 lines.

a. Batchian. Presented by W. W. Saunders, Esq.

Var.? *Fœm. Nigra, fusiformis, cinereo tomentosa; antennæ corpore multo breviores; prothorax luteo late marginatus; abdomen luteum, strigis ventralibus lateralibus transversis apiceque piceis; coxæ et trochanteres lutea; corium striga costali fasciaque exteriore luteis; membrana apicem versus cinerea.*

Female. Black, fusiform, with cinereous tomentum. Rostrum piceous, extending to the hind coxæ. Antennæ much shorter than the body; first joint much longer than the head; second rather shorter than the first; third much shorter than the second; fourth as long as the first. Prothorax broad, bordered with luteous in front and along each side; a slight transverse furrow at somewhat in front of the middle. Abdomen luteous; transverse streaks along each side beneath and tips piceous. Coxæ and trochanters luteous. Corium with a luteous costal streak, which extends nearly half the length from the base; an exterior luteous band. Membrane cinereous about the tip. Length of the body 6½ lines.

a. Sula. Presented by W. W. Saunders, Esq.

8. PHYSOPELTA FIMBRIATA.

fimbriata, *Stal, Berl. Ent. Zeit.* vii. 392.
Timor.

9. PHYSOPELTA SCHLANBUSCHII.

————— ———, *Stoll, Pun.* f. 273—Cimex Slanbuschii, *Fabr. Mant. Ins.* ii. 299—Lygæus Schlanbuschii, *Fabr. Ent. Syst.* iv. 155—Lygæus Schlanbuschii, *Fabr. Syst. Rhyn.* 222 — Pyrrhocoris Slanbuschi, *H.-Sch. Wanz. Ins.* ix. 179—Pyrrhocoris Schlangenbuschii, *Burm. Handb. Ent.* ii. 286—Physopelta Schlanbuschii, *Stal, Hem. Fabr.* i. 80.

a—c. Hong Kong. Presented by J. C. Bowring, Esq.
d. ————?

Var.? Mas et fœm. *Rufa, fusiformis; rostrum coxas posticas attingens; antennæ nigræ, corporis dimidio paullo longiores; prothorax maculis duabus anticis fasciaque postica abbreviata interrupta nigris; scutellum, tibiæ, tarsi, membranaque nigra; ventris latera nigro transverse quadristrigata; corium apud marginem interiorem nigro strigatum.*

Male and female. Red, fusiform, thickly and minutely punctured. Head triangular. Eyes piceous, not prominent. Rostrum black at the tip, extending to the hind coxæ. Antennæ black, a little more than half the length of the body; first joint hardly longer than the head; second a little longer than the first; third a little shorter than the first; fourth as long as the second. Prothorax with a slight transverse middle furrow; a small black spot on each side of the fore part, and an abbreviated and interrupted black band on the hind part. Scutellum black. Abdomen beneath with four transverse black streaks on each side. Fore femora incrassated, with three subapical spines beneath; tibiæ and tarsi black. Corium of the fore wings with a broad irregular black streak along the interior border; membrane black. Length of the body 6½ lines.

a, b. Penang. Presented by J. C. Bowring, Esq.

10. Physopelta pilosa.

pilosa, *Stal, Ofv. K. V. Ak. Forh.* 1870, 665.

Philippine Isles.

11. Physopelta apicalis.

Mas et fœm. Niger, fusiformis; caput subtus rufum; rostrum coxas posticas attingens; antennæ basi rufæ, corpore paullo breviores; prothorax transverse tenuiter sulcatus; abdomen rufo marginatum; pedes tibiis anticis femoribusque apice rufis, tibiis basi rufis, tarsis piceis; corium rufum, striga basali fascia apiceque nigris.

Male and female. Black, fusiform. Head red beneath. Rostrum extending to the hind coxæ; first joint red. Antennæ a little shorter than the body; first joint twice the length of the head, red at the tip; second much shorter than the first; third much shorter than the second; fourth much shorter than the third. Prothorax with a slight transverse middle furrow; disk black. Abdomen beneath with a red stripe; hind borders of the second and third segments much curved on each side. Connexivum red. Femora and fore tibiæ with red tips; tibiæ red at the base; tarsi piceous; fore femora with a row of spines beneath. Corium red, with a black basal streak, with a black band, which does not extend to the inner border, and with a black tip. Length of the body 8—9 lines.

a—d. Hindostan. From the Entomological Society's collection.

12. Physopelta plana.

Mas. Rufus, fusiformis; rostrum coxas intermedias attingens; antennæ nigræ, corporis dimidio longiores; prothorax transverse bisulcatus; scutellum piceum; segmenta pectoralia et ventralia nigro marginata; corium nigro unistrigatum et unimaculatum; membrana cinerea.

Male. Red, fusiform. Rostrum extending to the middle coxæ; tip black. Antennæ black, very much shorter than the body; first joint red at the base; second as long as the first; third very much shorter than the second. Prothorax with two transverse abbreviated furrows, the anterior one curved; fore part hardly convex; sides hardly reflexed. Scutellum piceous. Fore borders of the pectoral and ventral segments black. Corium with a broad black streak parallel to the interior border and with a black subapical spot. Membrane cinereous. Length of the body 11 lines.

a, b. Hindostan. From Mr. Children's collection.

13. Physopelta bimaculata.

Mas. Fulva, fusiformis; rostrum coxas posticas attingens; antennæ nigræ, corpore multo breviores, articulo 1o rufo apice nigro; dorsi abdominalis discus niger; corium macula subapicali nigra; membrana cinerea.

Male. Tawny, fusiform. Rostrum extending to the hind coxæ, black at the tip. Antennæ black, much shorter than the body; first joint red, black at the tip, longer than the head; second a little longer than the first. Prothorax with a slight transverse middle furrow and with a more slight curved transverse furrow near the fore border. Dorsum of the abdomen with a black disk. Corium with a round black subapical spot. Membrane and hind wings cinereous. Length of body 8½ lines.

a. Hindostan. From Mr. Stevens' collection.

14. Physopelta immunis.

Mas. Testacea, fusiformis; rostrum rufum, coxas posticas attingens; antennæ nigræ, corpore multo breviores, articulo 1o basi fulvo; prothorax sulcis duobus transversis nigris abbreviatis; segmenta pectoralia et ventralia suturis nigris; pedes rufi, tibiis piceis; corium nigro unimaculatum et uniguttatum; membrana pallide cinerea.

Male. Testaceous, fusiform. Rostrum red, extending to the hind coxæ; tip black. Antennæ black, much shorter than the body; first joint tawny at the base, a little longer than the head; second as long as the first; third very much shorter than the second; fourth a little shorter than the second. Prothorax with two slight black abbreviated transverse furrows; one a little before the middle, the other curved, near the fore border. Sutures of the pectoral and ventral segments black. Legs red; femora and tibiæ spinulose; tibiæ piceous. Corium with a black dot near the end of the interior border, and with a black spot near the end of the exterior border. Membrane pale cinereous. Length of the body 10—11 lines.

a, b. Siam. Presented by W. W. Saunders, Esq.
c. Cambodia. From M. Mouhot's collection.

Genus 10. ECTATOPS.

Ectatops, *Serv. Hist. Hem.* 273—Bislectum. Ectatops, *Schönh. Mant. Sec. Curc.* 1847, 19.

   A. Prothorax with strongly-marked transverse furrows; sides distinctly concise and reflexed.
   a. Corium with no marks.
   * Prothorax entirely black.
   † Membrane pale, with a black mark.   -   - ophthalmicus.
   †† Membrane pale, black-bordered.   -   -   - adustus.

** Prothorax red; hind part of disk sometimes black.
† Scutellum black.
‡ Hind disk of the prothorax black. - - - limbatus.
‡‡ Prothorax wholly red. - - - - distinctus.
‡‡‡ Fore part of the prothorax black. - - - ventralis.
‡‡‡‡ Prothorax black, with red borders. - - - largoides.
† Scutellum red. - · - - - rubiaceus.
b. Corium with marks.
* Abdomen blackish beneath.
† Membrane brown, with pale veins. - - gracilicornis.
†† Membrane brown, with a whitish border. - - tenuicornis.
††† Membrane wholly black. - - - - coloratus.
** Abdomen black, luteous-bordered. - - - subjectus.
*** Abdomen beneath reddish or yellowish.
† Head red. - - - - - - erythromelas.
†† Head black, ochraceous at the base. - - seminiger.
c. Corinm partly pale-bordered. - - - amabilis.
d. Corium with a red costal streak. - - - ruficosta.
B. Prothorax with no transverse furrow, or with a slight
    one; sides scarcely concise, hardly elevated
a. Head not elongated. Sides of the prothorax straight. lateralis.
b. Head elongated. Sides of the prothorax sinuated. - obscurus.

1. ECTATOPS LIMBATUS.

limbatus, *Serv. Hist. Hem.* 273; *De Vuill. A. S. E. F.* 4me *Sér.* iv. 144;
    Stal, Berl. Ent. Zeit. vii. 396; *K. Sv. Vet. Ak. Handl.* 1870, 105.

Hindostan. Java.

2. ECTATOPS RUBIACEUS.

rubiaceus, *Serv. Hist. Hem.* 273; *De Vuill. A. S. E. F.* 4me *Sér.* iv. 144;
    Stal, Berl. Ent. Zeit. vii. 396; *K. Sv. Vet. Ak. Handl.* 1870, 105.

a. Java.
b. Malacca. Presented by W. W. Saunders, Esq.
c, d. Singapore. Presented by W. W. Saunders, Esq.
e—j. Sarawak. Presented by W. W. Saunders, Esq.
k. Sarawak. From Mr. Wallace's collection.

3. ECTATOPS OPHTHALMICUS.

Pyrrhocoris ophthalmicus, *Burm. Handb. Ent.* ii. 284—Ectatops ophthal-
    micus, *De Vuill. A. S. E. F.* 4me *Sér.* iv. 143; *Stal, Berl. Ent. Zeit.*
    vii. 397; *K. Sv. Vet. Ak. Handl.* 1870, 106.

Java.

4. ECTATOPS LATERALIS.

lateralis, *De Vuill. A. S. E. F.* 4me *Sér.* iv. 144; *Stal, K. Sv. Vet. Ak.*
    Handl. 1870, 106.

Silhet.

5. ECTATOPS DISTINCTUS.

distinctus, *De Vuill. A. S. E. F.* 4*me Sér.* iv. 144 ; *Stal, K. Sv. Vet. Ak. Handl.* 1870, 106.

Silbet.

6. ECTATOPS OBSCURUS.

obscurus, *De Vuill. A. S. E. F.* 4*me Sér.* iv. 144 ; *Stal, K. Sv. Vet. Ak. Handl.* 1870, 106.

Malacca.

7. ECTATOPS ERYTHROMELAS.

erythromelas, *Stal, Berl. Ent. Zeit.* vii. 396 ; *K. Sv. Vet. Ak. Handl.* 1870, 105.

Cambodia.

8. ECTATOPS SEMINIGER.

seminiger, *Stal, Berl. Ent.* vii. 397 ; *K. Sv. Vet. Ak. Handl.* 1870, 105
  Ofv. K. V. Ak. Forh. 1870, 666.

Philippine Isles.

9. ECTATOPS GRACILICORNIS.

gracilicornis, *Stal, Berl. Ent. Zeit.* vii. 396 ; *K. Sv. Vet. Ak. Handl.* 1870
  105.

a—c. New Guinea.   Presented by W. W. Saunders, Esq.
d—f. Mysol.   Presented by W. W. Saunders, Esq.
g, h. Ké.   Presented by W. W. Saunders, Esq.
i. Aru.   Presented by W. W. Saunders, Esq.

10. ECTATOPS FUSCUS.

fuscus, *Stal, Ofv. K. V. Ak. Forh.* 1870, 667.

Philippine Isles.

11. ECTATOPS RUBENS.

rubens, *Stal, K. Sv. Vet. Ak. Handl.* 1870, 105 ; *Ofv. K. V. Ak. Forh.* 1870, 667.

Philippine Isles.

12. ECTATOPS LARGOIDES.

Mas.   *Niger, fusiformis; caput rufum ; rostrum segmenti ventralis 1i marginem posticum attingens ; antennæ basi rufæ ; prothorax rufo marginatus ; venter luteus.*

Male. Black, fusiform. Head red. Eyes lightly petiolated. Rostrum extending to the hind border of the first ventral segment. First joint of the antennæ a little longer than the head, red at the base. Prothorax bordered with red ; a strongly-marked transverse furrow at one-third of the length from the fore border. Abdomen luteous beneath. Fore femora slightly incrassated with two spines beneath towards the tips. Length of the body 7 lines.

a. Siam. Presented by W. W. Saunders, Esq.

13. ECTATOPS ADUSTUS.

Mas. *Niger, fusiformis, cinereo tomentosus; rostrum segmenti 1i ventralis marginem posticum attingens; antennæ corporis dimidio longiores; scutellum apice luteum; femora postica basi testacea; corium striga discali lutea; membrana albida, fusco marginata.*

Male. Black, fusiform, coarsely punctured above, with cinereous tomentum. Head elongate-triangular. Eyes very prominent, slightly ascending. Rostrum extending to the hind border of the first ventral segment. Antennæ rather more than half the length of the body; first joint hardly as long as the head; second much shorter than the first; third a little shorter than the second; fourth almost as long as the first. Prothorax with a well-defined transverse furrow at a little in front of the middle. Scutellum luteous at the tip. Fore femora slightly incrassated, with two minute teeth near the tips beneath; hind femora testaceous towards the base. Corium of the fore wings with a luteous streak in the disk beyond the middle; membrane whitish, with a narrow dark brown border. Length of the body 5½ lines.

a. Singapore. Presented by W. W. Saunders, Esq.

14. ECTATOPS TENUICORNIS.

Fœm. *Niger, fusiformis, cinereo tomentosus; rostrum coxas posticas attingens; antennæ gracillimæ, corporis dimidio æquilongæ, articulo 4o crassiore basi testaceo; prothorax fascia postica angusta testacea; abdominis stigmata atra; femora antica incrassata; corium testaceum, nigro punctatum unimaculatum et uniguttatum; membrana fusca, albida venosa et marginata.*

Female. Black, fusiform, finely punctured, with cinereous tomentum. Head triangular. Eyes piceous, very prominent. Rostrum extending to the hind coxæ. Antennæ very slender, about half the length of the body; first joint longer than the head; second a little more than half the length of the first; third much shorter than the second; fourth testaceous towards the base, thicker than the preceding joints, and a little longer than the second. Prothorax with a slight transverse furrow at somewhat before the middle, and with a narrow pale testaceous band near the hind border. Abdomen with deep black stigmata. Fore femora incrassated, minutely spinous beneath. Corium of the fore wings testaceous, with black punctures, with a black spot in the disk beyond the middle, and with a black apical dot; membrane brown, with a whitish border and with whitish-bordered veins. Length of the body 5 lines.

a. Singapore. Presented by W. W. Saunders, Esq.

15. Ectatops amabilis.

Mas. *Niger, fere linearis; capitis lobus intermedius luteus; rostrum coxas posticas attingens, articulo 4o luteo; antennæ basi luteæ, corpore multo breviores; prothorax postice ochraceus, lateribus albis; segmenta pectoralia albo marginata; abdominis dorsum ochraceum; venter flavo trifasciatus; pedes lutei; corium ochraceum, lineis duabus albis nigro marginatis; membrana nigra, basi apicem alba.*

Male. Black, smooth, shining, nearly linear. Head triangular; middle lobe luteous. Eyes livid, large, prominent. Rostrum extending to the hind coxæ; fourth joint luteous. Antennæ much shorter than the body; first joint luteous towards the base, very much longer than the head; second very much shorter than the first; third very much shorter than the second. Prothorax with a well-defined transverse middle furrow, ochraceous and very finely punctured between the furrow and the hind border; sides white. Pectoral segments bordered with white. Abdomen ochraceous above; under side with three broad yellow bands. Legs luteous; fore femora hardly incrassated. Corium ochraceous, with a white black-bordered line along the costa, and another along the flexure; membrane black, white at the base and at the tip. Length of the body 5 lines.

c. Aru. Presented by W. W. Saunders, Esq.

16. Ectatops subjectus.

Mas. *Niger, fusiformis, subtus cinereo tomentosus; rostrum rufum, coxas posticas attingens; antennæ corpore multo breviores, articulis 2o et 3o basi 4oque dimidio basali albidis; prothorax postice ochraceus, nigro punctatus; abdominis dorsum, pedes et corium lutea.*

Male. Black, fusiform, punctured, with cinereous tomentum beneath. Head triangular. Eyes lurid, very prominent. Rostrum red, extending to the hind coxæ. Antennæ much shorter than the body; first joint as long as the head; second and third whitish at the base; second very much shorter than the first; third shorter than the second; fourth whitish for nearly half the length from the base, as long as the second. Prothorax with a well-defined transverse middle furrow; hind part ochraceous, black-punctured. Dorsum, connexivum and tip of the abdomen, legs and corium luteous. Length of the body 5 lines.

a. Celebes. From Mr. Wallace's collection.

17. Ectatops venustus.

Mas. *Niger, fusiformis; rostrum segmenti 1i ventralis marginem posticum attingens; antennæ corpore breviores, articuli 4i dimidium basale album; prothorax antice albo postice luteo marginatus; venter luteo marginatus et late vittatus; pedes femoribus anticis subincrassatis, femoribus tarsisque quatuor posterioribus basi albis; corium costa fasciaque subapicali albis. Var. β.—Prothoracis margo anticus albus; antennarum articulus 4us omnino niger; ventris vitta indeterminata.*

Male. Black, fusiform, shining, very finely punctured. Head elongate triangular; middle lobe slightly prominent. Eyes very prominent. Rostrum extending a little beyond the fore border of the second ventral segment. Antennæ somewhat shorter than the body; first joint white at the base, a little shorter than the head; second a little shorter than the first; third rather shorter than the second; fourth longer than the first, sometimes white for full half the length from the base. Prothorax narrower in front, with a well-defined transverse middle furrow, narrowly bordered in front and on each side, and more broadly bordered behind with white, which hue on the hind border and on the adjoining part of each side is more or less tinged with luteous. Abdomen beneath narrowly bordered with whitish along each side and behind; under side with a broad and sometimes much abbreviated luteous stripe; hind border of the apical segment red. Four posterior femora white towards the base; fore femora slightly incrassated, with two teeth beneath near the tips. Middle tibiæ brown, sometimes whitish; fore tibiæ whitish. Tarsi brown; first joint whitish except at the tip. Corium with a whitish line, which extends from the base nearly to a white band near the tip. Membrane blackish. Length of the body 4—5 lines.

a. New Guinea. Presented by W. W. Saunders, Esq.
b. New Guinea. From Mr. Wallace's collection.

18. ECTATOPS COLORATUS.

Mas. *Niger, subfusiformis, cinereo subtomentosus; rostrum segmenti 2i ventralis marginem posticum attingens; antennæ corpore sat breviores, articulis 2o 3o 4oque basi albis, 3o subclavato; prothorax margine lateribusque anticis spatioque postico luteis; scutellum apice luteum; corium luteo submarginatum, apice album; membrana atra.*

Male. Black, subfusiform, with slight cinereous tomentum. Head triangular. Eyes very prominent, slightly ascending. Rostrum extending to the hind border of the second ventral segment. Antennæ somewhat shorter than the body; first joint a little longer than the head; second and third white at the base; second shorter than the first; third subclavate, shorter than the second; fourth white towards the base, a little longer than the third. Prothorax luteous and largely punctured between the hind border and a slight transverse furrow, which is rather before the middle; fore border and sides of the fore part very narrowly luteous. Scutellum luteous at the tip. Fore femora slightly incrassated, with two minute subapical teeth beneath; knees luteous. Corium of the fore wings luteous towards the base, white towards the tip, luteous also along the middle part of the costa and of the interior border; membrane deep black. Length of the body 6 lines.

a. Ceram. Presented by W. W. Saunders, Esq.

19. ECTATOPS RUFICOSTA.

Mas. *Niger, fusiformis; rostrum segmenti 1i ventralis marginem posticum attingens; antennæ gracillimæ, corpore breviores, articulo 4o basi fulvo; prothorax sulco transverso valde determinato, lateribus valde reflexis; abdomen rufum; pedes graciles, tibiis tarsisque quatuor anterioribus piceis; corium villa costali apicem versus dilatata; membrana obscure fusca.*

Male. Black, fusiform. Head shining; middle lobe prominent. Eyes livid, very prominent. Rostrum extending to the hind border of the first ventral segment. Antennæ very slender, rather shorter than the body; first joint very much longer than the head; second much shorter than the first; third shorter than the second; fourth longer than the second, shorter than the first, tawny at the base. Prothorax with a strongly-marked transverse furrow; sides much reflexed. Abdomen red. Four anterior tibiæ and tarsi piceous. Corium with a red costal stripe, which is dilated towards the tip. Membrane dark brown. Length of the body 5 lines.

a. New Guinea. Presented by W. W. Saunders, Esq.

Div. 2.

Euscopus, *Stal, K. Sv. Vet. Ak. Handl.* 1870, 106.

20. ECTATOPS RUFIPES.

Euscopus rufipes, *Stal, K. Sv. Vet. Ak. Handl.* 1870, 106.

Java.

Genus 11. LARGUS.

Largus, *Hahn, Wanz. Ins.* i. 13. *Burm. Handb. Ent.* ii. 281. *Blanch. Serv. Hist. Hem.* 273—Euryophthalmus, *De Lap.*

Most of the following species of Largus may be thus distinguished.

A. Femora not incrassated; hind femora unarmed.
  a. Membrane wholly black.
    * Legs wholly black.
    † Corium yellowish. -   -   -   - convivus.
    †† Corium red. -   -   -   - bicolor, rufipennis.
    ††† Corium black, with white lines. -   - divisus.
    ** Femora yellowish red. -   -   -   - socius.
  b. Membrane brown. -   -   -   - longulus.
  c. Membrane cinereous, with darker veins.
    * Prothorax partly pale.
    † Pale band of prothorax very broad.
    ‡ Corium black or blackish.
    § Hind border of pectus pale.
    × Veins of the membrane not black. -   - lineola.
    × × Veins of the membrane black.   balteatus.
    §§ Hind border of pectus black. -   - cinctiventris.
    ‡‡ Corium luteous.
    § Corium not spotted.
    × Abdomen with deep yellow bands. -   - trochanterus.
    × × Abdomen beneath with narrow black bands. - nigricollis.
    §§ Corium with two black spots. -   -   - anticus.

†† Pale band of prothorax rather narrow. - fascialis.
††† Pale band of prothorax very narrow.
‡ Corium not striped.
§ Femora luteous towards the base.
✕ Membrane brown.
o Antennæ stout. - - - - succinctus.
oo Antennæ slender. - - - - cinctus.
✕✕ Membrane white. - - - - concisus.
§§ Femora wholly luteous. - - - pulverulentus.
§§§ Femora black. - - - - obtusus.
‡‡ Corium striped. - - - - torridus.
** Prothorax black, with a pale border. - - bipustulatus.
*** Prothorax pale, with two black spots. - - discolor.
B. Femora somewhat incrassated, all with spines beneath.
a. Abdomen beneath with no white spots. - crassipes.
b. Abdomen beneath with three white spots on each side. - - - - fatidicus.

Prof. Stal considers that L. lineola is distinct from L. humilis.

He makes the following arrangement of the Largi, with which he includes Acinocoris and Lecadra.

A. Femoribus posticis inermibus.
a. Thorace basi hemelytrorum latitudine subæquali, lateribus haud ampliatis.
* Oculis longiuscule stylatis, stylo distincte sursum vergente; capite lobo antico thorace latitudine subæquali vel fere latiore; hemelytris medio haud vel vix ampliatis. lunatus.
** Oculis brevissime stylatis; capite lobo antico thoracis angustiore vel latitudine subæquali; thorace basi capite duplo vel fere duplo latiore; elytris medio ampliatis.
† Parte tota basali metasthetii pone impressionem transversam sita eburnea vel flavo-testacea.
‡ Segmento basali ventris eburneo margine basali nigro; limbo abdominis unicolore, immaculato; thorace anticis maculis duabus flavescentibus vel testaceis notato. humilis, balteatus, trochanterus.
‡‡ Segmento basali ventris nigro; limbo abdominis maculis flavo-testaceis notato; thorace anterius utrinque flavo-testaceo limbato. fasciatus.
†† Parte basali metasthetii concolore nigro, interdum in piceum vergente, angulis posticis raro eburneis.
‡ Ventre nigro.
§ Limbo exteriore corii flavescente vel testaceo vel rufo.
✕ Femoribus totis vel ad partem nigris. rufipennis, discolor, cinctiventris, longulus, convivus, cinctus, varians, succinctus.
✕✕ Femoribus totis testaceis vel flavo-testaceis. bipustulatus, socius.
§§ Hemelytris nigris, limbo exteriore concolore. morio, tristis.
‡‡ Ventre fere toto eburneo vel flavescente. xanthomelas, sellatus.

b. Thorace retrorsum sensim maxime ampliato, basi hemelytrorum
multo latiore, parte ampliata pone medium leviter reflexa. abdominalis.

B. Femoribus posterioribus subtus apicem versus spinulis armatis,
posticis apud marem incrassatis. lineola, crassipes, fatidicus.

1. LARGUS SUCCINCTUS.

Cimex rubrocinctus, *Deg. Ins.* iii. 339, pl. 34, f. 19—Cimex succinctus,
Linn. Cent. Ins. 17; *Syst. Nat.* i. 2, 727; *Amæn. Acad.* vi. 400.
Fabr. Syst. Ent. 723; *Sp. Ins.* ii. 369; *Mant. Ius.* ii. 303. *Gmel.*
ed. Syst. Nat. i. 2175. *Goeze. Ent. Beitr.* ii. 214. *Tigny, Hist. Ins.*
iv. 280—Lygæus succinctus, *Fabr. Ent. Syst.* iv. 170; *Syst. Rhyn.*
233—Largus succinctus, *Burm. Handb. Ent.* ii. 283. *H.-Sch. Wanz.*
Ins. vi. 78, pl. 206, f. 648. *Stal, Hem. Fabr.* i. 80; *K. Sv. Vet. Ak.*
Handl. 1870, 94.

a. United States. Presented by E. Doubleday, Esq.
b. United States. Presented by W. W. Saunders, Esq.
c. Mexico. Presented by E. P. Coffin, Esq.
d. Mexico. From M. Hartweg's collection.
e. Orizaba. From M. Sallé's collection.
f. North America. Presented by F. Walker, Esq.
g. California. From M. Hartweg's collection.

2. LARGUS LINEOLA.

Cimex lineola, *Linn. Syst. Nat.* i. 2, 722. *Gmel. ed. Syst. Nat.* ii. 2155—
Cimex punctatus, *Deg. Ins.* iii. 337, pl. 34, f. 17—18. *Stoll, Pun.*
f. 19, 145—Lygæus gibbus, *Fabr. Syst. Rhyn.* 227—Cimex humilis,
Drury, Ins. iii. 65, pl. 46, f. 3. *Stoll, Pun.* pl. 37, f. 265—Lygæus
mutilus, *Perty, Del. An. Art.* 173, pl. 34, f. 9—Euryophthalmus
puncticollis, *De Lap. Ess. Hem.* 38—Largus lineola, *Stal, Hem. Fabr.*
i. 80; *K. Sv. Vet. Ak. Handl.* 1870, 96—Largus humilis, *Hahn,*
Wanz. Ins. i. 13, pl. 2, f. 6. *Burm. Handb. Ent.* ii. 282. *Stal, K.*
Sv. Vet. Ak. Handl. 1870, 92.

a—d. Rio Janeiro. Presented by the Rev. H. Clark.
e, f. Petropolis. Presented by the Rev. H. Clark.
g—i. Presented by J. Gray, Esq.
j—m. Constancia. Presented by the Rev. H. Clark.
n—r. Tejuca. Presented by the Rev. H. Clark.
s. British Guiana. Presented by Sir R. Schomburgk.
t. Santarem. Presented by W. W. Saunders, Esq.
u. Santarem. From Mr. Bates' collection.
v. Brazil. Presented by J. P. G. Smith, Esq.
w, x. ———? From Mr. Children's collection.

3. LARGUS BALTEATUS.

balteatus, *Stal. K. Sv. Vet. Ak. Handl.* 1870, 92.
Bolivia.

4. LARGUS CINCTUS.

cinctus, *H.-Sch. Wanz. Ins.* vii. 6, pl. 218, f. 683. *Stal, Stett. Ent. Zeit.* xxiii. 315; *K. Sv. Vet. Ak. Handl.* 1870, 94—Capsus succinctus, Var. *a., Say, New Harm. Ind.* 1831 ; *Works ed Lec.* i. 338.

a. United States. Presented by E. Doubleday, Esq.
b. United States. Presented by W. W. Saunders, Esq.
c. Orizaba. From M. Sallé's collection.
d. Columbia. From M. Goudot's collection.
e. ———?

5. LARGUS CONVIVUS.

convivus, *Stal, Ofv. K. V. Ak. Forh.* 1861, 196; *K. Sv. Vet. Ak. Handl.* 1870, 94.

Mexico.

6. LARGUS LONGULUS.

longulus, *Stal, Ofv. K. V. Ak. Forh.* 1861, 196; *K. Sv. Vet. Ak. Handl.* 1870, 94.

Mexico.

7. LARGUS VARIANS.

varians, *Stal, K. Sv. Vet. Ak. Handl.* 1870, 94—cinctus, var.?

Bogota. New Granada.

8. LARGUS RUFIPENNIS.

Euryophthalmus rufipennis, *De Lap. Ess. Hem.* 39—Largus rufipennis, *Burm. Handb. Ent.* ii. 283. *Blanch. Hist. Ins.* 128. *Serv. Hist. Hem.* 274. *Stal, K. Sv. Vet. Ak. Handl.* 1870, 93—Largus sanguinipennis, *H.-Sch. Wanz. Ins.* ix. 182—Largus bicolor, *H.-Sch. Wans. Ins.* vii. 7 —Largus marginicollis, *Stal, Ofv. K. V. Ak. Forh.* 1855, 186.

a—c. Rio Janeiro. Presented by the Rev. H. Clark.
d—j. Constantia. Presented by the Rev. H. Clark.
k—n. Constancia. Presented by J. Gray, Esq.
o—r. ——— ? From Mr. Children's collection.

9. LARGUS BIPUSTULATUS.

bipustulatus, *Stal, Ofv. K. V. Ak. Forh.* 1861, 196; *K. Sv. Vet. Ak. Handl.* 1870, 95.

Mexico.

10. LARGUS SOCIUS.

socius, *Stal, Ofv. K. V. Ak. Forh.* 1861, 197; *K. Sv. Vet. Ak. Handl.* 1870, 95.

Mexico.

12. LARGUS FATIDICUS.

fatidicus, *Stal, Ofv. K. V. Ak. Forh.* 1861, 197; *K. Sv. Vet. Ak. Handl.*
  1870, 96.
Brazil.

13. LARGUS MORIO.

morio, *Stal, Ofv. K. V. Ak. Forh.* 1855, 186; *K. Sv. Vet. Ak. Handl.*
  1870, 95.
New Granada.

14. LARGUS TRISTIS.

tristis, *Stal, K. Sv. Vet. Ak. Handl.* 1870, 95.
New Granada.

15. LARGUS CINCTIVENTRIS.

cinctiventris, *Stal. Rio Jan. Hem.* 44; *K. Sv. Vet. Ak. Handl.* 1870, 94.
Rio Janeiro.

16. LARGUS TROCHANTERUS.

trochanterus, *Sgnt. A. S. E. F.* 4me *Sér.* ii. 583.   *Stal, K. Sv. Vet. Ak.*
  Handl. 1870, 93—lineola, *Serv. Hist. Hem.* 274.
Brazil.   Peru.

17. LARGUS FASCIATUS.

fasciatus, *Blanch. Orb. Voy. Amer. Merid.* vi. 220, pl. 30, f. 6.   *Stal, K.*
  Sv. Vet. Ak. Handl. 1870, 93.
St. Jago d'Estrella.

18. LARGUS CRASSIPES.

crassipes, *Stal, Ofv. K. V. Ak. Forh.* 1861, 197; *K. Sv. Vet. Ak. Handl.*
  1870, 96.
Surinam.

19. LARGUS XANTHOMELAS.

Lygæus xanthomelas, *Perty, Del. An. Art. Bras.* 172, pl. 34, f. 6—Largus
  pulchellus, *Blanch. Hist. Ins.* 128—Largus xanthomelas, *Stal, K. Sv.*
  Vet. Ak. Handl. 1870, 95.
North Brazil.

20. LARGUS SELLATUS.

Lygæus (Largus) sellatus, *Guér. Sagra, Hist. Cuba, Ins.* 401—Largus
  sellatus, *Stal, K. Sv. Vet. Ak. Handl.* 1870, 95.
Cuba.

21. Largus concisus.

Fœm. *Saturate rufus, ellipticus, robustus, capite rostro antennis prothorace antico pectore abdomine pedibusque nigris; rostrum coxas intermedias attingens; antennæ corporis dimidio paullo longiores; scutellum et corium basi nigra; corium margine exteriore nigro; membrana albida.* Var. β.—*Pectus ferrugineum; venter strigis utrinque transversis lanceolatis apiceque flavis.*

Female. Deep red, stout, elliptical, largely punctured. Head, rostrum, antennæ, fore division of prothorax, pectus, abdomen and legs black. Head triangular. Eyes very prominent. Rostrum extending to the middle coxæ. Antennæ a little more than half the length of the body; first joint longer than the head; second hardly more than half the length of the first; third shorter than the second; fourth shorter than the first. Prothorax with a distinct transverse furrow near the fore border. Scutellum and corium black at the base. Legs stout; fore femora with three spines beneath. Corium with a black exterior border; membrane whitish. *Var.* β.—Pectus ferruginous. Abdomen beneath with transverse lanceolate yellow streaks on each side; tip also yellow. Length of the body 6 lines.

a. Amazon Region. Presented by W. W. Saunders, Esq.
b. Demerara. Presented by F. Moore, Esq.
c. Brazil. Presented by W. W. Saunders, Esq.
d, e. Rio Negro. From Mr. Wallace's collection.

22. Largus pulverulentus.

Mas. *Niger, fusiformis, robustus, cano tomentosus; rostrum coxas intermedias attingens; prothorax margine lateribusque posticis flavis; pedes rufi, tibiis tarsisque nigris; corium flavum, nigro punctatum et unimaculatum; membrana nigra.*

Male. Black, fusiform, stout, with hoary tomentum. Head conical. Eyes piceous, very prominent. Rostrum extending to the middle coxæ. First joint of the antennæ about twice the length of the head. Prothorax with a very slight transverse furrow at somewhat behind the middle; hind part of each side and hind border pale yellow. Legs red, stout; tibiæ and tarsi black; fore femora beneath with two black subapical spines. Corium pale yellow, thickly punctured with black; a large transverse black spot in the disk beyond the middle. Membrane black. Length of the body 6½—7½ lines.

a. Orizaba. From M. Sallé's collection.

23. Largus obtusus.

Fœm. *Niger, fusiformis, subtus cinereo tomentosus; rostrum coxas intermedias attingens; antennæ corporis dimidio paullo longiores; prothorax luteo marginatus; scutellum apice luteum; corii costa lutea; membrana fusca, nigro venosa.*

Female. Black, fusiform, very finely punctured, with cinereous tomentum beneath. Head triangular. Eyes very prominent. Rostrum extending to the middle coxæ. Antennæ a little more than half the length of the body; first joint nearly twice the length of the head; second a little more than half the length of the first; third much shorter than the second; fourth much longer than the second. Prothorax bordered with pale luteous; a slight transverse middle furrow; hind half roughly punctured. Scutellum luteous at the tip. Fore femora beneath with two subapical spines. Corium rather roughly punctured, with a luteous costa; membrane brown, with black veins. Length of the body 8 lines.

a. Orizaba.   From M. Sallé's collection.

24. LARGUS TORRIDUS.

Fœm. *Niger, fusiformis, crassus; rostrum coxas posticas attingens; antennæ corporis dimidio longiores; prothorax rufo tenuiter marginatus; venter nigro rufus; pedes breviusculi, femoribus quatuor posterioribus basi apiceque rufis; corium rufo trivittatum et unimaculatum; membrana alba, fusco marginata.*

Female. Black, fusiform, thick, very finely punctured. Head triangular. Eyes livid, very prominent. Rostrum extending to the hind coxæ. Antennæ more than half the length of the body; first joint very much longer than the head; second less than half the length of the first; third much shorter than the second; fourth as long as the second and third together. Prothorax narrowly bordered with dark red; a very slight transverse middle furrow. Abdomen beneath blackish red. Legs stout, rather short; fore femora beneath with two minute subapical spines; four posterior femora dark red at the base and at the tips. Corium with three red stripes and with a red spot between the two exterior stripes, of which one is costal; membrane white, bordered with brown. Length of the body 6½ lines.

a, b. Columbia.   From M. Goudot's collection.

25. LARGUS ANTICUS.

Mas et fœm. *Niger, fusiformis, cinereo tomentosus, prothorace postico corioque ochraceis nigro punctatis; rostrum coxas intermedias attingens; antennæ corpore multo breviores; venter flavus, segmentis nigro fasciatis; corium nigro unimaculatum; membrana nigra.*

Male and female. Black, fusiform, with cinereous tomentum. Head, fore part of the prothorax and under side shining. Head triangular. Eyes piceous, very prominent. Rostrum extending to the middle coxæ. Antennæ much shorter than the body; first joint very much longer than the head; second hardly half the length of the first; third much shorter than the second; fourth much longer than the second. Prothorax with a distinct transverse middle furrow; hind half and corium ochraceous, with black punctures. Abdomen beneath yellow, with a black band on the fore border of each segment. Fore femora incrassated, with four spines beneath towards the tips. Corium with a black spot near the interior angle; membrane black. Length of the body 6—7 lines.

a, b. Villa Nova.   From Mr. Bates' collection.
c, d. Amazon River.   Presented by W. W. Saunders, Esq.

26. LARGUS DIVISUS.

Fœm.   *Ater, fusiformis, subtilissime punctatus; caput glabrum, bisul-*
catum ; oculi valde prominuli ; rostrum coxas intermedias attingens ;
antennæ corporis dimidio longiores ; prothorax transverse subsulcatus,
fascia postica rufa utrinque dilatata et elongata ; metapectus et
segmentum 1um *ventrale transverse eburneo lineata ; fasciæ tres*
posteriores ventrales eburneæ latæ interruptæ abbreviatæ ; femora
antica subtus spinosa ; corium costa margineque interiore tenuiter
eburneo marginata, striga obliqua discali strigaque clavata marginali
eburneis.

Female.   Deep black, fusiform, very finely punctured. Head small,
triangular, smooth, with two longitudinal furrows between the eyes, which
are very prominent. Rostrum extending to the middle coxæ. Antennæ
slender, rather more than half the length of the body; first joint nearly
twice the length of the head ; second hardly half the length of the first;
third much shorter than the second; fourth nearly as long as the first.
Prothorax convex, with a slight transverse furrow at a little in front of the
middle ; a red band along the hind border, dilated on each side, where it
extends to the transverse furrow. Scutellum with a white line towards the
tip. Metapectus and basal ventral segment with a transverse ivory-white
line on each side. Second, third and fifth ventral segments with an
interrupted abbreviated and very broad ivory-white band on each. Fore
femora with spines of various size beneath. Corium with the costa and
the interior border very narrowly ivory-white ; an ivory-white streak along
the interior border, widening towards the tip ; another ivory-white streak
from the base of the costa nearly to the inner angle. Length of the body
6 lines. ·

a. Tunantins.   From Mr. Bates' collection.

Country unknown.
27. LARGUS SEXGUTTATUS.
sexguttatus, *H.-Sch. Wanz. Ins.* ix. 181.   ·

28. LARGUS PALLIDICORNIS.
pallidicornis, *H.-Sch. Wanz. Ins.* ix. 181.

Genus 12. LECADRA.
Lecadra, *Sgnt. A. S. E. F.* 4me *Sér.* ii. 582.

1. LECADRA ABDOMINALIS.
Lecadra abdominalis, *Sgnt. A. S. E. F.* 4me *Sér.* ii. 582, pl. 15, f. 3—
    Largus abdominalis, *Stal, K. Sv. Vet. Ak. Handl.* 1870, 96.
Peru.

a. Para.   Presented by W. W. Saunders, Esq.
b. Para.   From Mr. Bates' collection.
c. Tapayos.   From Mr. Bates' collection.
d. Ega.   From Mr. Bates' collection.

Genus 13. FIBRENUS.

Fibrenus, *Stal*, *K. Ofv. Vet. Ak. Forh.* 1861, 195.

1. FIBRENUS GLOBICOLLIS.

Largus globicollis, *Burm. Handb. Ent.* ii. 282—Fibrenus gibbicollis, *Stal*, *Ofv. K. V. Ak. Forh.* 1861, 198; *K. Sv. Vet. Ak. Handl.* 1870, 96.

Para.

a. Columbia. From M. Jurgenson's collection.
b—e. Oajaca. From M. Sallé's collection.

Genus 14. ARHAPHE.

Arhaphe, *H.-Sch. Wanz. Ins.* ix. 183.

1. ARHAPHE CAROLINA.

Carolina, *H.-Sch. Wanz. Ins.* ix. 183, pl. 315, f. 968.

Mexico.

a. Georgia. Presented by E. Doubleday, Esq.

2. ARHAPHE CICINDELOIDES.

Mas et fœm. *Niger, fusiformis, hirsutus, cinereo tomentosus ; caput globosum, atro uninotatum ; rostrum coxas intermedias fere attingens ; antennæ corporis dimidio longiores ; prothorax convexus ; pedes sub-hirsuti ; corium albo bimaculatum.*

Male and female. Black, fusiform, hirsute, with cinereous tomentum. Head globose, with a deep black mark on the vertex. Eyes piceous, prominent. Rostrum extending nearly to the middle coxæ. Antennæ more than half the length of the body ; first joint a little shorter than the head ; second a little shorter than the first ; third much shorter than the second ; fourth a little longer than the first. Prothorax convex, without any transverse furrow. Legs slightly hirsute ; fore femora slightly dilated. Fore wings extending to two-thirds of the length of the abdomen, each with two large white spots. Length of the body 4 lines.

a—c. Oajaca, Mexico. From M. Sallé's collection.
d. Mexico. From Mr. Glennie's collection.

Genus 15. ASTEMMA.

Astemma, *St. Farg. et Serv. Enc. Méth.* x. 323. *Stal, K. Sv. Vet. Ak. Handl.* 1870, 91—Acinocoris, *Serv. Hist. Hem.* 275.

1. ASTEMMA CORNUTA.

Astemma cornuta, *St. Farg. et. Serv. Enc. Méth.* x. 323—Acinocoris cornutus, *Serv. Hist. Hem.* 275.

Cayenne.

2. ASTEMMA STYLOPHTHALMA.

stylophthalmum, *Stal, K. Sv. Vet. Ak. Handl.* 1870, 91.

North Brazil.

Var.? Mas. *Atra, fere linearis ; rostrum coxas intermedias attingens ; antennæ corporis dimidio longiores ; prothorax antice, pectus et venter cano tomentosa ; venter flavo latissime bifasciatus ; corium læte ochraceum.*

Male. Deep black, nearly linear, very finely punctured. Head triangular; front nearly perpendicular. Eyes slightly petiolated, obliquely ascending. Rostrum extending to the middle coxæ. Antennæ more than half the length of the body; first joint longer than the head; second about half the length of the first; third much shorter than the second; fourth rather shorter than the first. Prothorax with hoary tomentum in front of the strongly-marked transverse middle furrow. Pectus and under side of the abdomen with hoary tomentum, the latter with two very broad yellow bands in the middle. Fore femora slightly incrassated, with two minute subapical spines beneath. Corium very bright orange. Length of the body 6 lines.

a. Ega. From Mr. Bates' collection.

Genus 16. ACINOCORIS.

Acinocoris, *Hahn, Wanz. Ins.* ii. 113. *Serv. Hist. Hem.* 274.

1. ACINOCORIS LUNATUS.

Cimex lunatus, *Fabr. Mant. Ins.* ii. 302—Cimex lunaris, *Gmel. ed. Syst. Nat.* i. 2, 178—Lygæus lunulatus, *Fabr. Ent. Syst.* iv. 167; *Syst. Rhyn.* 232—Lygæus calidus, *Fabr. Syst. Rhyn.* 230—Lygæus lunulatus, *Fabr. Syst. Rhyn. Index,* 16—Acinocoris calidus, *Hahn, Wanz. Ins.* ii. 114, pl. 64, f. 194. *Blanch. Hist. Ins.* 128—Largus lunulatus, *Burm. Handb. Ent.* ii. 282. *Blanch. Hist. Ins.* 227—Largus interruptus, *H.-Sch. Wanz. Ins.* ix. 181, pl. 317, f. 978—Acinocoris lunatus, *Stal, Hem. Fabr.* i. 81—Largus lunatus, *Stal, K. Sv. Vet. Ak. Handl.* 1870, 92.

a. Para. Presented by R. Graham, Esq.
b. Para. Presented by J. P. G. Smith, Esq.

2. ACINOCORIS LUNULATUS.

Lygæus lunulatus, *Fabr. Syst. Rhyn.* 232—Largus lunulatus, *Burm. Handb. Ent.* ii. 282—Acinocoris interruptus, *H.-Sch. Wanz. Ins.* ix. 118, pl. 317, f. 978.

Cayenne.

a. Santarem. From Mr. Bates' collection.
b—d. Tapayos. From Mr. Bates' collection.
e. Peru. From M. Degand's collection.

Note.—Cimex oculi-cancri, *Deg. Ins.* iii. 343, pl. 34, f. 24. *Tign. Hist. Nat. Ins.* iv. 274, may belong to this genus.

3. Acinocoris bilineatus.

Mas. *Niger, ellipticus, cinereo tomentosus; rostrum segmentum 1um
ventrale attingens; antennæ corpore multo breviores; prothorax
lateribus margineque postico luteis; venter luteo maculariter
bivittatus; pedes lutei, femoribus nigro notatis, tibiis basi nigris;
corium costa pallide testacea, linea discali lutea; membrana cinereo
tenuiter marginata.*

Male. Black, elliptical, punctured, with cinereous tomentum. Head
triangular. Eyes very prominent. Rostrum extending to the first ventral
segment. Antennæ much shorter than the body; first joint very much
longer than the head; second about half the length of the first; third
much shorter than the second; fourth much longer than the second. Prothorax with a strongly-marked middle furrow; sides and hind border
luteous. Legs luteous; femora with some black marks; tibiæ black at the
base. Corium with a pale testaceous costa and with an oblique luteous
line in the disk. Membrane with a narrow cinereous border. Length of
the body 4½ lines.

a. Demerara. Presented by F. Moore, Esq.

4. Acinocoris includens.

Mas. *Niger, subfusiformis; rostrum coxas intermedias attingens;
antennæ corporis dimidio longiores; prothorax postice flavo trimaculatus; venter vittis duabus macularibus flavis; femora tibiæque
flava, illa nigro vittata; corium flavo bivittatum; membrana albido
tenuiter marginata.*

Male. Black, subfusiform, somewhat roughly punctured. Head triangular. Eyes very prominent. Rostrum extending to the middle coxæ.
Antennæ rather more than half the length of the body; first joint longer
than the head; second very much shorter than the first; third much
shorter than the second; fourth a little shorter than the first. Prothorax
with a strongly-defined transverse furrow at a little in front of the middle;
a yellow spot on the hind border and one on each hind angle. Abdomen
beneath with two macular yellow stripes. Femora and tibiæ yellow, the
former striped with black; fore femora beneath with a minute subapical
spine. Corium with a yellow costal stripe and a curved yellow stripe in
the disk, joining the costa at each end; membrane with a narrow whitish
border. Length of the body 4 lines.

a, b. Cuenca. From Mr. Fraser's collection.

Genus 17. THERANEIS.

Theraneis, *Spin. Ess. Hém.* 181—Theraneis et Stenomacra, *Stal, K. Sv.
Vet. Ak. Handl.* 1870, 97.

1. Theraneis scapha.

Lygæus scapha, *Perty, Del. An. Art. Bras.* 172, pl, 34, f. 8—Largus
incisus, *H.-Sch. Wanz. Ins.* ix. 182, pl. 317, f. 981—Stenomacra
scapha, *Stal, K. Sv. Vet. Ak. Handl.* 1870, 97.

a. Brazil.

2. THERANEIS MARGINELLUS.

Largus marginellus, *H.-Seh. Wanz. Ins.* ix. 182, pl. 317, f. 982 (parallelus)—Theraneis marginella, *Stal, Ent. Zeit. Stett.* xxiii. 315—Stenomacra marginella, *Stal, K. Sv. Vet. Ak. Handl.* 1870, 98.

a, b. Mexico. Presented by E. P. Coffin, Esq.
c. Mexico. From Mr. Glennie's collection.
d. Oajaca. From M. Sallé's collection.
e. Orizaba. From M. Sallé's collection.

3. THERANEIS CLIENS.

Theraneis cliens, *Stal, Stett. Ent. Zeit.* xxiii. 314—Stenomacra cliens, *Stal, K. Sv. Vet. Ak. Handl.* 1870, 98.
Mexico.

4. THERANEIS LIMBATIPENNIS.

Theraneis limbatipennis, *Stal, Rio Jan. Hem.* 45—Stenomacra limbatipennis, *Stal, K. Sv. Vet. Ak. Handl.* 1870, 98.
Rio Janeiro.

5. THERANEIS VITTATA.

vittata, *Spin. Ess. Hém.* 181; *Stal, K. Sv. Vet. Ak. Handl.* 1870, 97.
Rio Janeiro.

Var.? *Atra; oculi valde prominuli; antennæ corporis dimidio longiores; prothorax sulco transverso medio bene determinato sulculoque postico longitudinali canis; femora antica trispinosa, subincrassata; corium vitta costali coccinea stramineo marginata.*

Male. Deep black, narrow, linear. Head triangular, flat between the eyes, which are very prominent. Antennæ rather more than half the length of the body; first joint a little longer than the head; second hardly more than half the length of the first; third much shorter than the second; fourth a little shorter than the first. Prothorax with a well-defined hoary transverse middle furrow and with a slight longitudinal furrow from thence to the hind border. Fore femora slightly incrassated, with three spines beneath. Corium with a crimson costal stripe, which is irregularly straw-colour along the inner side. Length of the body 4½—5 lines.

a, b. Petropolis. Presented by the Rev. H. Clark.

6· THERANEIS CONSTRICTA.

constricta, *Stal, K. Sv. Vet. Ak. Handl.* 1870, 97.
New Granada. Bogota.

7. THERANEIS FERRUGINEA.

ferruginea, *Mayr, Verh. Zool. Bot. Ges. Wien.* xv. 436. *Stal, K. Sv. Vet. Ak. Handl.* 1870, 97.

Brazil.

Prof. Stal considers that the following species are probably Pyrrhocoridæ :—

Capsus ocreatus, *Say, New Harm. Ind.* 1831; *Works ed. Lec.* i. 338.
Georgia.

Lygæus fuscipennis, *Guér. Voy. Coquille Ins.* 178, pl. 12, f. 14.
Port Praslin.

Lygæus Woodlarkianus, *Mrtz. Ann. Sci. Phys. et Nat.* 2me Sér. vii. 105.
Woodlark. Physopelta famelica ?

Lygæus cruciatus, *Mrtz. Ann. Sci. Phys. et Nat.* 2me Sér. vii. 106.
Woodlark.

Lygæus Fabricii, *Mrtz. Ann. Sci. Phys. et Nat.* 2me Sér. vii. 106.
Woodlark.

Lygæus violaceus, *Mrtz. Ann. Sci. Phys. et Nat.* 2me Sér. vii. 107.
Woodlark.

BICELLULI.

Bicelluli, *Serv. Hist. Hem.* 275—Bicelluli et Unicelluli, *Sgnt. A. S. E. F.* 3me Sér. vi. 499.

Fam. 1. CAPSIDÆ.

Mirides et Capsides, *Serv. Hist. Hem.* 277, 278—Astemmites, p. *De Lap.* —Capsini, *Burm. Handb. Ent.* ii. 263. *Kol.*—Capsidæ, *Westw.*— Capsina, *Flor.*—Phytoceridæ, *Fieb. Eur. Hem.* 237—Capsina, Unicelluli and Bicelluli, *Dougl. and Scott, Hem.* 27.

The following synopsis of the genera into which the European species of this family have been divided is translated from Fieber's Eur. Hem.

A. Membrane with single almost half-round areolets.
  a. Rostrum long, extending to the middle of the
    metapectus. - - - - - MONALOCORIS.
  b. Rostrum short, extending to the end of the meso-
    pectus. - - - - - - BRYOCORIS.

B. Membrane with double generally elongate triangular rectilinear or curvilinear areolets.
a. First joint of the hind tarsi twice or thrice longer
than the second joint.
* Prothorax as broad as long, reverse-trapeziform,
much straightened hindward. Mesothorax uncovered. - - - - - MYRMECORIS.
** Prothorax elongate-trapeziform. Mesothorax
covered.
† Head with rounded sides. - - - PITHANUS.
†† Head with parallel sides.
‡ Prothorax with no rim on the fore border.
§ Cheek-plates short, almost semicircular.
✕ Tip of the vertex truncated, even.
o Rostrum extending to the second ventral segment. MIRIS.
oo Rostrum extending to the hind border of the
mesopectus. - - - - - BRACHYTROPIS.
✕✕ Tip of the vertex conical.
⟶ First joint of the rostrum longer than the under
side of the head. - - - - NOTOSTIRA.
⟶⟶ First joint of the rostrum as long as the under
side of the head. - - - - LOBOSTETHUS.
§§ Cheek-plates long, linear.
✕ First joint of the rostrum longer than the under
side of the head. - - - - MEGALOCEROEA.
✕✕ First joint of the rostrum as long as the under
side of the head. - - - - TRIGONOTYLUS.
‡‡ Prothorax with a rim on the fore border.
§ Sides of the prothorax foliaceous or sharp-bordered.
✕ Tip of the vertex forming a prominent cone. - ACETROPIS.
✕✕ Tip of the vertex rounded.
o Eyes half-round. Rostrum extending to the hind
border of the metapectus. - - - LEPTOPTERNA.
oo Eyes oval, almost reniform. Rostrum extending
to the hind border of the mesopectus. - - TERATOCORIS.
§§ Sides of the prothorax truncated or rounded.
✕ First joint of the hind tarsi not thicker than the
other joints.
o Second joint of the antennæ clavate. - CREMNOCEPHALUS.
oo Second joint of the antennæ linear. - ONCOGNATHUS.
✕✕ First joint of the tarsi much thicker than the two
following joints.
o Second joint of the antennæ clavate. First joint
of the hind tarsi cylindrical. - - - ALLOEOTOMUS.
oo Joints of the antennæ cylindrical. First joint of
the hind tarsi clavate. - - - - PACHYPTERNA.
b. First joint of the hind tarsi shorter than the
second joint, or equal to it in length.
* Head above transversely oval or triangular. Eyes
contiguous, or nearly so.
† Prothorax in front with a rim or swelling.
‡ Neck arched, with no transverse border or rim.

§ Fore border of the prothorax with a narrow rim.

✕ Callus of the front more or less pointed at the base, or extended into an almost right angle.

  o First joint of the hind tarsi as long as the second.

   → Areolets of the membrane half-round. .   - CAMPTOBROCHYS.

 →→ Areolets of the membrane elongate-triangular.

   ++ Tip of the scutellum obtusely short-conical   - CONOMETOPUS.

 ++++ Tip of the scutellum curved downward.

  ᴕ Head vertical, almost quadranguler.   -   - MEGACOELUM.

 ᴕᴕ̄ Head quadrangular, somewhat horizontal.   - HOMODEMUS.

  oo First joint of the hind tarsi shorter than the second.

   → Rostrum short, extending to the mesopectus.   - BRACHYCOLEUS.

 →→ Rostrum extending to the hind coxæ or beyond them.

   ++ Prothorax trapeziform, with straight sides.

  ᴕ Rostrum extending to the second ventral segment.   CALOCORIS.

 ᴕᴕ Rostrum extending to the middle of the abdomen beneath.

   + Head with almost parallel sides.   MIRIDIUS.

  ++ Head with angular sides.   -   -   - PHYTOCORIS.

 ++++ Prothorax elongate-trapeziform, with excavated sides.

  ᴕ Second joint of the antennæ cylindrical.

   + Head horizontal.   -   -   -   - ALLOEONOTUS.

  ++ Head tumid.   -   -   -   -   - HALLODAPUS.

 ᴕᴕ Second joint of the antennæ clavate above.   - CLOSTEROTOMUS.

✕✕ Callus of the front curved.

  o Second joint of the antennæ cylindrical.   - PYCNOPTERNA.

  oo Second joint of the antennæ clavate.

   → Prothorax rectangular, almost cylindrical.   - GRYLLOCORIS.

 →→ Prothorax trapeziform.

   ++ All the joints of the rostrum stout.   -   - RHOPALOTOMUS.

 ++++ Second, third and fourth joints of the rostrum slender.   -   -   -   -   - CAPSUS.

§§ Fore border of the prothorax with a broad rim or swelling.

✕ Transverse callus of the prothorax limited by the side borders.

  o Sides of the prothorax foliaceous in front.   - LOPUS.

  oo Sides of the prothorax obtuse, almost tumid in front.   -   -   -   -   - HORISTUS.

✕✕ Transverse callus of the prothorax extending to the sides of the pectus.

  o Areolet of the membrane elongate-triangular.   - DYONCUS.

  oo Areolet of the membrane curved.   -   - CAMPYLONEURA.

 ‡‡ Neck with a complete transverse border, or with short rims by the eyes.

  § Transverse border of the neck only visible by the eyes.

✕ Third joint of the tarsi longer than the second.   - DICHROOSCYTUS.

✕✕ Third joint of the tarsi almost shorter than the second.   -   -   -   -   - LIOCORIS.

§§ Transverse border of the neck complete.
✕ Cuneus short-triangular, curvilinear, little longer than the breadth of its base.
o Head thick. - - - - - CHARAGOCHILUS.
oo Head slender, elongated.
-+ Rostrum extending to the hind border of the metapectus. - - - - - POLYMERUS.
-+-+ Rostrum extending to the second or third ventral segment. - - - - - CYPHODEMA.
✕✕ Cuneus rectilinear, almost twice as long as the breadth of its base.
§§ Fore border of the prothorax with a broad rim.
✕ Callus of the front conspicuous, extending in a sharp angle of the front.
o Head thick. - - - - - TYLONOTUS.
oo Head less thick, somewhat elongated.
-+ Rostrum extending to the second or to the third ventral segment. - - - - LYGUS.
-+-+ Rostrum extending to the middle of the metapectus. - - - - - - PŒCILOSCYTUS.
✕✕ Callus of the front passing with a curve into the vertex.
o Third joint of the hind tarsi somewhat longer than the second. - - - - - HADRODEMA.
oo Third joint of the hind tarsi somewhat shorter than the second. - - - - ORTHOPS.
†† Prothorax with no rim in front.
‡ Areolets of the wings without hooks.
§ Head transverse.
✕ Callus of the front extending almost into a quadrant. - - - - - STIPHROSOMA.
✕✕ Callus of the front compressed, more or less prominent.
o Cheeks elevated into ridges along the eyes. - HALTICUS.
oo Cheeks not elevated into ridges along the eyes.
-+ Head remarkably straightened behind the eyes.
++ Third joint of the hind tarsi longer than the second. - - - - - CYLLOCORIS.
++++ Third joint of the hind tarsi shorter thad the second. - - - - - GLOBICEPS.
-+-+ Head not straightened behind the eyes.
++ Prothorax hell-shaped.
ᴖ Head not thick, compressed between the eyes. - MECOMMA.
ᴖᴖ Head thick.
⊙ Prothorax short-bell-shaped. - - - CYRTORHINUS.
⊙⊙ Prothorax elongate-bell-shaped. - - AETORHINUS.
++++ Prothorax transverse- or elongate-trapeziform.
ᴖ Second joint of the antennæ cylindrical, or slightly clavate.
⊙ Rostrum extending to the hind border of the mesopectus.
= Second joint of the hind tarsi longer than the third. PACHYLOPS.

== Second joint of the hind tarsi shorter than the
   second. - - - - -   CAMPTOTYLUS.
⊙⊙ Rostrum extending to the hind border of the
   metapectus, or beyond it.
= Vertex not dilaled. -   -   -   -   LOXOPS.
== Vertex dilated.
 V Callus of the front stout, curved and prominent. -   LITOCORIS.
VV Callus of the front vertical, equally broad.
 ⇌ Head above transverse. -   -   -   -   XENOCORIS.
⇌⇌ Head above with nearly equal sides.   -   -   ORTHOTYLUS.
ↄↄ Second joint of the antennæ very clavate or
   thickly cylindrical in the male, thick clavate-
   spindle-shaped or compressed in the female.
 ⊙ Rostrum extending to the second ventral segment.   HETEROTOMA.
⊙⊙ Rostrum extending to the hind border of the
   metapectus. -   -   -   -   HETEROCORDYLUS.
§§ Head elongated.
 ✕ Second joint of the hind tarsi longer than the
   others. -   -   -   -   -   ORTHOCEPHALUS.
✕✕ Third joint of the hind tarsi longer than the
   others. -   -   -   -   -   LABOPS.
‡‡ Areolets of the wings with hooks.
 § Second joint of the antennæ spindle-shaped.   -   ATRACTOTOMUS.
§§ Second joint of the antennæ clavate.
 ✕ Callus of the front proceeding from an almost
   right angle.
  o Xyphus swollen.
 ⊷ Second joint of the antennæ shorter than the
   third. -   -   -   -   -   -   HARPOCERA.
⊷⊷ Second joint of the antennæ longer than the
   third. -   -   -   -   -   MEGALODACTYLUS.
oo Xyphus deepened.
 ⊷ Eyes elongated or oval.
++ Rostrum extending to the hind border of the
   metapectus. -   -   -   -   -   ONCOTYLUS.
++++ Rostrum extending to the third or fourth ventral
   segment. -   -   -   -   -   CONOSTETHUS.
                                                 ANOTEROPS.
⊷⊷ Eyes round. -   -   -   -   -
✕✕ Callus of the front proceeding from a more or less
   acute angle or from a curve.
  o Xyphus swollen.
 ⊷ Callus of the front proceeding from a more or less
   acute angle.
++ Face oblique. -   -   -   -   -   TINICEPHALUS.
++++ Face vertical.
 ↄ Face bent in the fore part.   -   -   -   TRAGISCUS.
ↄↄ Face straight in the fore part.
 ⊙ Prothorax trapeziform.
 = Throat conspicuous.
 V First joint of the rostrum as long as the under
   side of the head. -   -   -   -   BRACHYORTHRUM.
VV First joint of the rostrum extending to the middle
   of the xyphus.

⇌ Second joint of the antennæ thick and cylindrical in the male, clavate in the female. - - CRIOCORIS.

⇌⇌ Antennæ of the male and female alike.

Λ Second joint of the hind tarsi longer than the third. - - - - - - PLAGIOGNATHUS.

ΛΛ Second joint of the hind tarsi as long as the third. - - - - - - APOCREMNUS.

== Throat hardly conspicuous.

V Antennæ long. - - - - - PSALLUS.

VV Antennæ short.

⇌ Hind tarsi short, stout. - - - - STHENARUS.

⇌⇌ Hind tarsi long, slender. - - - AGALLIASTES.

☉☉ Prothorax elongate-bell-shaped.

== Second joint of the hind tarsi longer than the third. - - - - - - MALTHACUS.

=== Second joint of the hind tarsi as long as the third. AUCHENOCREPIS.

→→ Callus of the front proceeding from a curve.

++ Neck angular. - - - - - CAMARONOTUS.

++++ Neck rounded. - - - - PHYLUS.

oo Xyphus even or deepened.

→ Xyphus even.

++ Face vertical. - - - - - GNOSTUS.

++++ Face oblique.

ഗ Second joint of the hind tarsi as long as the third. HOPLOMACHUS.

ഗഗ Second joint of the hind tarsi longer than the third.

☉ Rostrum extending to the third or fourth ventral segment. - - - - - PACHYXYPHUS.

☉☉ Rostrum extending to the first ventral segment. - PLACOCHILUS.

→→ Xyphus deepened.

++ Throat about half as long as the head. - - MACROTYLUS.

++++ Throat as long as the head beneath.

ഗ Head elongated. - - - - AMBLYTYLUS.

ഗഗ Head hardly elongated. - - - - MACROCOLEUS.

* Head elongated or oval. Eyes apart.

† Head elongated. - - - - MACROLOPHUS.

†† Head oval or nearly round.

‡ Edge of the throat armed with two little teeth. - ODONTOPLATYS.

‡‡ Edge of the throat unarmed.

§ Prothorax trapeziform. - - - - MALACOCORIS.

§§ Prothorax elongated.

✕ Base of the callus and insertion of the antennæ between the eyes. - - - - CYRTOPELTIS.

✕✕ Base of the callus and insertion of the antennæ lower.

o Head oblique. - - - - - SYSTELLONOTUS.

oo Head vertical.

→ Head above little longer than its breadth behind. BRACHYCERÆA.

→→ Head above nearly twice longer than its breadth behind. - - - - - DICYPHUS.

The following divisions of this family are recorded in the Catalogue, but are not included in the preceding Synopsis:—

Acropelta, Microsynamma, Grypocoris, Bothynotus, Stethoconus, Exæretus, Tytthus, Dasycytus, Aspicelus, Perideris, Zygimus, Plagiorhamma, Platycranus, Myrmecophyes, Liops, Stenoparia.

This family, as regards the British species, is divided by Douglas and Scott into fifteen families, which are here enumerated with their respective genera.

       Div. 1. Unicelluli.
1. Bryocoridæ.—Bryocoris. Monolocoris.
       Div. 2. Bicelluli.
2. Pithanidæ.—Pithanus.
3. Miridæ.—Miris. Acetropis. Lopomorpha.
4. Phytocoridæ.—Miridius. Phytocoris.
5. Deræocoridæ.—Deræocoris. Pantilius.
6. Litosomidæ.—Litosoma.
7. Phylidæ.—Ætorhinus. Sphyracephalus. Byrsoptera. Phylus.
8. Camaronotidæ.—Camaronotus.
9. Globicepidæ.—Globiceps.
10. Idolocoridæ.—Cyllocoris. Systellonotus. Campyloneura. Idolocoris.
    Macrolophus. Malacocoris.
11. Oncotylidæ.—Anoterops. Macrocoleus. Amblytylus. Tinicephalus.
    Oncotylus. Hoplomachus. Conostethus.
12. Psallidæ.—Plagiognathus. Apocremnus. Psallus. Sthenarus.
13. Capsidæ.—Neocoris. Agalliastes. Orthocephalus. Heterocordylus.
    Atractotomus. Heterotoma. Rhopalotomus. Capsus. Polymerus.
14. Lygidæ. — Charagochilus. Camptobrochys. Liocoris. Orthops.
    Lygus. Pœciloscytus.
15. Harpoceridæ.—Harpocera.
16. Myrmicocoridæ.—Myrmicocoris.
17. Lopidæ.—Lopus.
18. Dichrooscytidæ.—Dichrooscytus.
19. Halticocoridæ.—Halticocoris.
20. Stiphrosomidæ.—Stiphrosoma.

In this Catalogue the species of Capsidæ are nearly all included in the following genera.
    A. Fore tibiæ not dilated.
    a. Prothorax not gibbous in front.
    * Tarsi not broader towards the tips.
    † First joint of the tarsi not as long as the
        second and third together.
    ‡ Scutellum with no spine.
    § Prothorax not contracted in the middle.
    ✗ Membrane with two areolets.

o Prothorax with straight sides.   -   -   1. Miris.

oo Prothorax with reflexed sides.   -   -   2. Lopus.

ooo Prothorax with widened sides.

-+ First joint of the antennæ as long as the prothorax, or longer.   -   -   . -   3. Phytocoris.

-+-+ First joint of the antennæ shorter than the
prothorax, rarely nearly equal to it in
length.

++ Prothorax broader in front than behind.

++++ Prothorax not broader in front than behind.

ꞷ Prothorax long, with a distinct transverse
furrow. -   -   -   -   -   4. Cyllecoris.

ꞷꞷ Prothorax with no distinct transverse furrow.

+ Prothorax with a ring-shaped ridge on the
fore border. -   -   -   -   5. Capsus.

++ Prothorax with no ring-shaped ridge on the
fore border.

= Hind femora not or only slightly thickened
or compressed. -   -   -   - 6. Leptomerocoris.

== Hind femora in the female, at least, very
distinctly thickened or compressed. - 7. Eurymerocoris.

✕✕ Membrane with one areolet.

o Rostrum not extending to the tip of the
abdomen.

-+ Body short.

++ Eyes not very prominent.

ꞷ Rostrum extending to the middle of the
metapectus. -   -   -   -   8. Monolocoris.

ꞷꞷ Rostrum extending to the end of the metapectus. -   -   -   -   -   9. Bryocoris.

ꞷꞷꞷ Rostrum hardly extending beyond the hind
coxæ. -   -   -   -   -   10. Fulvius.

++++ Eyes very prominent. -   -   - 11. Anapus.

oo Rostrum extending to to the tip of the abdomen. -   -   -   -   - 13. Psilorhamphus.

-+-+ Body long, slender.   12. Myrmecoris. 14. Disphinctus.
15. Monalonion. 16. Eucerocoris. 17. Pachypetis.

§§ Prothorax contracted in the middle.   18. Sphinctothorax.

‡‡ Scutellum with an erect spine. 19. Helopeltis. 20. Herdonius.

‡‡ First joint of the tarsi about as long as the
second and third together.   ..   •   21. Valdasus.

** Tarsi broader towards the tips.

† Corium with no appendage.   -   - 22. Eccritotarsus.

†† Corium with an appendage.   -   -· 23. Sinervus.

b. Prothorax gibbous in front, and extending
over the head. -   -   -   - 24. Ambracius.

B. Fore tibiæ much dilated.   -   -   - 25. Hemicocnemis.

The four following genera are thus distinguished by Signoret.

A. Scutellum flat.

a. First joint of the antennæ short.   -   •   - Monalonion.

b. First joint of the antennæ long.   -   -   -   EUCEROCORIS.
B. Scutellum tumid.   -   -   -   -   -   PACHYPELTIS;
C. Scutellum with a spine.   -   -   -   -   HELOPELTIS.

Genus 1. MIRIS.

Miris, *Fabr. Syst. Rhyn.* 36; *Serv. Hist. Hem.* 277; *Wolff, Hahn, Fall.*
   Burm. Handb. Ent. ii.; *Blanch. Kirschb. Caps.* 29; *Fieb. Crit. Gen.*
   5; *Eur. Hem.* 62, 259; *Dougl. and Scott, Hem.* 282.

Europe.
Div. 1.
1. MIRIS LÆVIGATUS.

Cimex lævigatus, *Linn. Syst. Nat.* 730; *Faun. Suec.* 958; *Deg. Ins.* iii.
   192; *H.-Sch. Wanz. Ins.* iii. 43, pl. 85, f. 259; *Serv. Hist. Hem.*
   277; *Meyer, Dür. Caps.* 35; *Kol. Mel. Ent.* ii. 98; *Schill. Arb.* 52—
   Miris lævigatus, *Fabr. Syst. Rhyn.* 253; *Fall. Hem. Suec.* 130;
 . *Burm. Handb. Ent.* ii. 265; *Kirsch. Rhyn.* 33; *Fieb. Eur. Hem.*
   240; *Flor, Rhyn. Liv.* i. 425, ii. 608; *Dougl. and Scott, Hem.* 284—
   Miris virens, *Hahn, Wanz. Ins.* ii. 79, pl. 54, f. 165.

a—l. England.   From Mr. Stephens' collection.
m—o. Europe.   From Mr. Children's collection.

2. MIRIS SERICANS.

sericans, *Fieb. Eur. Hem.* 240.

Austria.

3. MIRIS HOLSATUS.

holsatus, *Fabr. Syst. Rhyn.* 254; *Fall. Hem. Suec.* 132; *Kirschb. Rhyn.*
   34; *Hahn, Wanz. Ins.* iii. 41, pl. 85; f. 256; *Zett. Ins. Lapp.* 280;
   Meyer, Dür. Caps. 36; *Kol. Mel. Ent.* ii. 99; *Schill. Arb.* 52; *Flor,*
   Rhyn. Liv. i. 427; *Sahlb. Geoc. Fen.* 86; *Fieb. Eur. Hem.* 241;
   Dougl. and Scott, Hem. 283.

Kamtschatka.

a—c. England.   From Mr. Stephens' collection.
d. England.   Presented by F. Walker, Esq.
e—g. England.
h—j. South France.
k, l. Europe.   From Mr. Children's collection.
m—o. Italy.   Presented by F. Walker, Esq.

Div. 2.

Brachytropis, *Fieb. Crit. Gen.* 6, pl. 6, f. 18; *Eur. Hem.* 62, 241.

4. MIRIS CALCARATUS.

calcaratus, *Fall. Hem. Suec.* i. 131; *Burm. Handb. Ent.* ii. 265; *H.-Sch.*
   Wanz. Ins. iii. 39; *Zett. Ins. Lapp.* 280; *Serv. Hist. Hem.* 278;
   Meyer, Dür. Caps. 34; *Schill. Arb.* 52; *Sahlb. Geoc. Fen.* 86;
   Kirschb. Rhyn. 33; *Flor, Rhyn. Liv.* i. 421; *Dougl. and Scott, Hem.*
   286—Miris dentatus, *Hahn, Wanz. Ins.* i. 15, pl. 2, f. 8—Brachytropis
   calcarata, *Fieb. Eur. Hem.* 241.

a--i. England. From Mr. Stephens' collection.
j. Isle of Wight. Presented by F. Walker, Esq.
k. Isle of Man. Presented by F. Walker, Esq.
l. Geneva. Presented by F. Walker, Esq.
m. Fontainbleau. Presented by F. Walker, Esq.
n, o. Europe. From Mr. Children's collection.
p. New York. Presented by E. Doubleday, Esq.
r. Nova Scotia. From Lieut. Redman's collection.
s. St. Martin's Falls, Albany River, Hudson's Bay. Presented by Dr.
Barnston.

Div. 3.

Notostira, *Fieb. Crit. Gen.* 7; *Eur. Hem.* 62, 241.

5. MIRIS ERRATICUS.

Cimex erraticus, *Linn. Syst. Nat.* ii. 731; *Faun. Suec.* 961—Miris horto-
rum, *Wolff, Icon. Cim.* f. 154—Miris erraticus, *Fall. Hem. Suec.* i.
132; *Hahn, Wanz. Ins.* ii. 78, pl. 54, f. 163, 164; *Burm. Handb.
Ent.* ii. 265; *H.-Sch. Wanz. Ins.* iii. 40; *Meyer Dür. Caps.* 34;
Serv. Hist. Hem. 278; *Kol. Mel. Ent.* ii. 96; *Schill. Arb.* 52; *Sahlb.
Geoc. Fen.* 87; *Kirsch. Rhyn.* 32; *Flor. Rhyn. Liv.* i. 431, ii. 608;
Dougl. and Scott, Hem. 287—Miris Caucasica, *Kol. Mel.* 97, pl. 11,
f. 25—Miris Tritici, *Curt. Brit. Ent.* xv. 701—Notostira erratica,
Fieb. Eur. Hem. 242.

Siberia.

a—d. England. From Mr. Stephens' collection.
e, f. England. Presented by F. Walker, Esq.
g—j. Europe.
k, l. Italy. Presented by F. Walker, Esq.

Div. 4.

Lobostethus, *Fieb. Crit. Gen.* 8, pl. 6, f. 19; *Eur. Hem.* 62, 242.

6. MIRIS VIRENS.

Cimex virens, *Linn. Syst. Nat.* i. 2, 102—Miris fulvus, *Fieb. Weit. Beit.*
i. 101—Miris lævigatus, *Hahn, Wanz. Ins.* ii. 76, pl. 53, f. 161—
Miris virens, *Kirsch. Rhyn.* 3—Lobostethus virens, *Fieb. Eur. Hem.*
243.

Siberia.

a—c. England. From Mr. Stephens' collection.
d, e. England.
f. Killarney, Presented by F. Walker, Esq.
g. Isle of Man. Presented by F. Walker, Esq.
h. Fontainbleau. Presented by F. Walker, Esq.
i—k. Europe. From Mr. Children's collection.

Div. 5.

Megaloceræa, *Fieb. Crit. Gen.* 9; *Eur. Hem.* 62, 243.

7. MIRIS LONGICORNIS.

longicornis, *Fall. Hem. Suec.* 129. *H.-Sch. Wanz. Ins.* iii. 43, pl. 85, f. 258. *Kirsch. Rhyn.* 32. *Meyer Dür. Caps.* 37. *Sahlb, Geoc. Fen.* 87. *Flor, Rhyn. Liv.* i. 434. *Dougl. and Scott, Hem.* 289—Miris megatoma, *Muls. Am. Soc. Linn.* 107—Megaloceræa longicornis, *Fieb. Eur. Hem.* 243.

a—d. England. From Mr. Stephens' collection.
e. Polish Ukraine. Presented by Dr. Dowler.
f—j. Italy. Presented by F. Walker, Esq.

Div. 6.

Trigonotylus, *Fieb. Crit. Gen.* 10, pl. 6, f. 20; *Eur. Hem.* 62, 243.

8. MIRIS RUFICORNIS.

ruficornis, *Fall. Hem. Suec.* i. 133. *Meyer Dür. Rhyn.* 37. *Kirsch. Rhyn.* 32. *H.-Sch. Wanz. Ins.* iii. 40. *Zett. Ins. Lapp.* 281. *Schill. Arb.* 52. *Sahlb. Geoc. Icon.* 87. *Flor, Rhyn. Liv.* i. 435; ii. 608—*Dougl. and Scott, Hem.* 290—Trigonotylus ruficornis, *Fieb. Eur. Hem.* 243—Miris pulchellus, *Hahn, Wanz. Ins.* ii. 119, pl. 66, f. 200.

Siberia.

a—l. England. From Mr. Stephens' collection.
m. N. Wales. Presented by F. Walker, Esq.
n. Europe. From Mr. Children's collection.
o, p. Italy. Presented by F. Walker, Esq.

Div. 7.

Acetropis, *Fieb. Crit. Gen.* ii. pl. 6, f. 2; *Eur. Hem.* 62, 244. *Dougl. and Scott, Hem.* 290—Lopomorphus, *Dougl. and Scott. Hem.* 293.

9. MIRIS CARINATUS.

Lopus carinatus, *H.-Sch. Wanz. Ins.* vi. 49, pl. 197, f. 609. *Kirschb. Rhyn.* 8—Acetropis carinatus, *Fieb. Eur. Hem.* 244—Lopomorphus carinatus, *Dougl. and Scott, Hem.* 294—Miris tricostata, *Costa.*

Europe.

10. MIRIS SETICULOSUS.

Acetropis seticulosa, *Pict. Mey. Fieb. Eur. Hem.* 244. *Dougl. and Scott, Hem.* 291.

England. Spain.

Div. 8.

Leptopterna, *Fieb. Crit. Gen.* 12, pl. 6, f. 3.   *Eur. Hem.* 63, 245—Lopomorphus, *Dougl. and Scott, Hem.* 293.

11. MIRIS DOLOBRATUS.

Cimex dolobratus, *Linn. Syst. Nat.* 730; *Faun. Suec.* 959.   *Scop. Ent·
Carn.* 135.   *Schr. Ins. Austr.* 285—Cimex lateralis, *Fabr. Gen. Ins·
Mant.* 300; *Sp. Ins.* ii. 373; *Mant. Ins.* ii. 306—Miris dolabratus,
Syst. Rhyn. 253.   *Hahn, Wanz. Ins.* ii. 75, pl. 53, f. 160.   *Fall. Hem.
Suec.* i. 128.   *Zett. Ins. Lapp.* 280.   *Schill. Arb.* 52.   *Flor, Rhyn.
Liv.* i. 437—Miris lateralis, *Fabr. Ent. Syst.* iv. 184 ; *Syst. Rhyn.*
254.   *Wolff, Icon. Cim.* f. 109—Miris ferrugatus, *Fall. Hem. Suec.*
129—Phytocoris dolabratus, *Burm. Handb. Ent.* ii. 267—Leptopterna
dolobrata, *Fieb. Eur. Hem.* 245—Lopus dolabratus, *H.-Sch. Wanz.
Ins.* iii. 45, pl. 86, f. 261, 362.   *Kirschb. Rhyn.* 36.   *Mey. Rhyn.* 38.
Sahlb. Geoc. Fen. 88—Lopus ferrugatus, *H.-Sch. Wanz. Ins.* iv. 46,
pl. 86, f. 263.   *Mey. Rhyn.* 39.   *Kirsch. Rhyn.* 12—Miris abbreviatus,
Wolff, Icon. Cim. f. 110—Lopomorphus dolabratus, *Dougl. and Scott,
Hem.* 297.

a—l. England.   From Mr. Stephens' collection.
m—p. England.
q, r. Suffolk.
s, t. Europe.   From Mr. Children's collection.
u. Switzerland.   Presented by Dr. Dowler.
v. Chamouni.   Presented by F. Walker, Esq.
z. South France.   Presented by F. Walker, Esq.
aa. Italy.   Presented by F. Walker, Esq.

Div. 9.

Teratocoris, *Fieb. Crit. Gen.* 13; *Eur. Hem.* 63, 245.

12. MIRIS ANTENNATUS.

Capsus antennatus, *Boh. Nya Suec. Hem.* 24—Teratocoris antennatus,
Fieb. Eur. Hem. 246.

Sweden.   Germany.

13. MIRIS NOTATUS.

Teratocoris antennatus, *Bär. Berl. Ent. Zeit.* 1859, 336, pl. 6, f. 9—
notatus, *Fieb. Eur. Hem.* 246.

Dalmatia.

14. MIRIS VIRIDIS.

Teratocoris viridis, *Dougl. and Scott, Ent. M. Mag.* iv. 46, pl. 1, f. 2.
Scotland.

15. Miris dorsalis.

Teratocoris dorsalis, *Fieb. Wien. Ent. Mon.* viii. 325.
Prague?

16. Miris Saundersi.

Teratocoris Saundersi, *Dougl. and Scott, Ent. M. Mag.* v. 260.
England.

Div. 10.

Cremnodes, *Fieb. Crit. Gen.* 14, pl. 6, f. 27—Cremnocephalus, *Fieb. Eur. Hem.* 63, 246.

17. Miris umbratilis.

Capsus umbratilis, *Fall. Hem. Suec.* 121—Cremnocephalus umbratilis, *Fieb. Eur. Hem.* 249.
Sweden.

Div. 11.

Miridius, *Fieb. Crit. Gen.* 25; *Eur. Hem.* 65, 257. *Dougl. and Scott, Hem.* 290.

18. Miris virgatus.

Miris virgatus, *Costa, Cent.* 1852—Miridius virgatus, *Fieb. Eur. Hem.* 258 —Miridius quadrivirgatus, *Dougl. and Scott, Hem.* 300.
England. Spain.
a, b. Italy. Presented by F. Walker, Esq.

North America.
19. Miris dorsalis.

dorsalis, *Say, Works ed Leconte,* ii. 348.
United States.

South America.
20. Miris insuavis.

insuavis, *Stal, Rio Jan. Hem.* 45.
Rio Janeiro.

21. Miris spurius.

spurius, *Stal, Eug. Resa,* 254.
Puna.

22. Miris scenicus.

scenicus, *Stal, Eug. Resa,* 254.
Rio Janeiro. Buenos Ayres.

23. Miris Dohrni.

Dohrni, *Stal, Eug. Resa*, 254.

Port Famine.

Genus 21. LOPUS.

Lopus, *Hahn, Wanz. Ins.* i. 143. *Kirschb. Caps.* 29. *Fieb. Grit. Gen.* 33 ;
Eur. Hem. 66, 266. *Dougl. and Scott, Hem.* 474.

Europe.
Div. 1.
1. Lopus cingulatus.

Cimex cingulatus, *Fabr. Mant. Ins.* ii. 307—Miris cingulatus, *Fabr. Ent.
Syst.* iv. 186; *Syst. Rhyn.* 255—Lopus albomarginatus, *Hahn, Wanz.
Ins.* i. 140, pl. 22, f. 72; *Fieb. Eur. Hem.* 267—Phytocoris albostriatus, *Klug, Burm. Handb. Ent.* ii. 271. *Mey. Rhyn.* 40—Lopus
albostriatus, *Kirschb. Caps.* 38.

a. Geneva.   Presented by M. Goureau.
b. Europe.   Presented by W. W. Saunders, Esq.

2. Lopus Mat.

Cimex Mat, *Rossi. Faun. Etr.* ii. 1346, pl. 7, f. 6—Phytocoris erythromelas, *Küst. Hahn, Wanz. Ins.* iii. 75, pl. 75, f. 231—Miris infuscatus? *Brullé, Exp. Mor.* 27—Lopus Mat, *Fieb. Eur. Hem.* 267.

South Europe.

3. Lopus Gothicus.

Cimex Gothicus, *Linn. Syst. Nat.* 2726; *Faun. Suec.* 966. *Scop. Ent.
Carn.* 131. *Schr. Ins. Austr.* 283. *Fabr. Mant. Ins.* ii. 305—
Lygæus Gothicus, *Wolff, Icon. Cim.* 33, pl. 4, f. 33—Capsus Gothicus,
Fabr. Syst. Rhyn. 244. *Panz. Faun. Germ.* 92. *Fall. Hem. Suec.* i.
117. *Flor, Rhyn. Liv.* i. 479—Lopus Gothicus, *Hahn, Wanz. Ins.* i.
12, pl. 2, f. 5. *Mey. Caps.* 465. *Kol. Mel. Ent.* ii. 100. *Kirschb.
Caps.* 37. *Fieb. Eur. Hem.* 267. *Dougl. and Scott, Hem.* 475.

Siberia.

a—j. England.   From Mr. Stephens' collection.
k, l. England.   Presented by F. Walker, Esq.
m—p. England.
q. Jersey.   Presented by the Entomological Club.
r. Polish Ukraine.   Presented by Dr. Dowler.
s, t. South France.   Presented by F. Walker, Esq.
u, v. Geneva.   Presented by M. Goureau.

4. Lopus superciliosus.

Cimex superciliosus, *Linn. Syst. Nat.* 2, 728—Lygæus albomarginatus,
Fabr. Ent. Syst. iv. 180—Capsus albomarginatus, *Fabr. Syst. Rhyn.*

245. *Fall. Hem. Suec.* i. 117. *Coq. Ill. Icon.* pl. 10, f. 2—Phytocoris
Gothicus, *Burm. Handb. Ent.* ii. 272 —Capsus Gothicus, var., *Flor,*
Rhyn. Liv. i. 480—Lopus superciliosus, *Douglas and Scott, Ent. M.*
Mag. iv. 51.
Europe.

6. LOPUS FLAVOMARGINATUS.

Cimex flavomarginatus, *Donov. Nat. Hist. Brit. Ins.* vii. 79, pl. 245—
Lopus Miles, *Dougl. and Scott, Hem.* 476—Gothicus, var. ?—Lopus
flavomarginatus, *Dougl. and Scott, Ent. M. Mag.* iv. 52.
England.

6. LOPUS SULCATUS.

sulcatus, *Pict. Mey. Fieb. Eur. Hem.* 268.
Spain.

7. LOPUS? LINEOLATUS.

Miris lineolatus, *Brullé, Exp. Mor.* 76, pl. 31, f. 6, 7—Lopus? lineolatus,
Fieb. Eur. Hem. 268.
Sicily.

8. LOPUS? CRUENTATUS.

Miris cruentatus, *Brullé, Exp. Mor.* 78, pl. 31, f. 8—Lopus? cruentatus,
Fieb. Eur. Hem. 268.
Sicily.

9. LOPUS SATYRISCUS.

satyriscus, *Scott, Stett. Ent. Zeit.* xii. 101.
Spain.

Div. 2.

Horistus, *Fieb. Eur. Hem.* 66, 268.

10. LOPUS RUBROSTRIATUS.

rubrostriatus, *H.-Sch. Wanz. Ins.* iii. 45, pl. 86, f. 260—Horistus rubrostriatus, *Fieb. Eur. Hem.* 268.
Spain. Dalmatia. Turkey.

Div. 3.

Pantilus, *Curt. Ent. Mag.* i. 197. *Dougl. and Scott, Hem.* 332—Conometopus, *Fieb. Crit. Gen.* 20, pl. 6, f. 1; *Eur. Hem.* 64, 249.

11. LOPUS TUNICATUS.

Cimex tunicatus, *Fabr. Mant. Ins.* ii. 203—Lygæus tunicatus, *Fabr. Ent.*
Syst. iv. 170; *Syst. Rhyn.* 233—Phytocoris tunicatus, *Fall. Hem.*
Suec. i. 85—Miris tunicatus, *Germ. Ahr. Faun. Ins. Eur.* 5, pl. 23—
Pantilus tunicatus, *Curt. Ent. Mag.* i. 197. *Dougl. and Scott, Hem.*

333—Lopus tunicatus, *Meyer*, *Caps.* 40. *Kirschb. Caps.* v. 37. *Flor, Rhyn. Liv.* i. 441—Conometopus tunicatus, *Fieb. Eur. Hem.* 249.

a—l. England. From Mr. Stephens' collection.
m—p. England. Presented by F. Walker, Esq.
q. Italy. Presented by F. Walker, Esq.
r. Europe. From Mr. Children's collection.

12. LOPUS PRASINUS.

Conometopus prasinus, *Fieb. Verh. Zool. Bot. Ges. Wien.* xx. 258.
Sarepta.

Div. 4.

Amblytylus, *Fieb. Crit. Gen.* 87, pl. 6, f. 22. *Eur. Hem.* 76, 318. *Dougl. and Scott, Hem.* 388.

13. LOPUS ALBIDUS.

Miris albidus, *Hahn, Wanz. Ins.* ii. 77, pl. 53, f. 162—Lopus albidus, *Kirschb. Caps.* 35—Amblytylus albidus, *Fieb. Eur. Hem.* 318.
Germany.

14. LOPUS BREVICOLLIS.

Amblytylus brevicollis, *Fieb. Crit. Gen.* 87, Sp. 35 ; *Eur. Hem.* 318.
South France. Corsica.

15. LOPUS LUNULA.

Amblytylus lunula, *Pict. Mey. Fieb. Eur. Hem.* 318.
South Spain.

16. LOPUS LONGIROSTRIS.

Amblytylus longirostris, *Pict. Mey. Fieb. Eur. Hem.* 319.
South Spain.

17. LOPUS NASUTUS.

Lopus nasutus, *Kirschb. Caps.* 10—Amblytylus nasutus, *Fieb. Eur. Hem.* 319.
Germany.

18. LOPUS JANI.

Amblytylus Jani, *Fieb. Crit. Gen.* 87, sp. 36 ; *Eur. Hem.* 319.
Italy.

19. LOPUS AFFINIS.

Amblytylus affinis, *Fieb. Wien. Ent. Mon.* viii. 332. *Dougl. and Scott, Hem.* 389, pl. 21, f. 3.

England. Germany.

20. LOPUS BICOLOR.

bicolor, *Fieb. Wien. Ent. Mon.* viii. 328.

Tauria.

Div. 7.

21. LOPUS CRUCIATUS.

cruciatus, *Sahlb. Geoc. Fen.* 89. *Fieb. Eur. Hem.* 391.

North Europe.

America.

22· LOPUS FILICORNIS.

Capsus filicornis, *Fabr. Syst. Rhyn.* 245.

South America.

23. LOPUS SULCATICORNIS.

sulcaticornis, *Stal, Rio Jan. Hem.* 46.

Rio Janeiro.

24. LOPUS RUFINASUS.

rufinasus, *Stal, Rio Jan. Hem.* 45.

Rio Janeiro.

25. LOPUS HAHNI.

Hahni, *Stal, Rio Jan. Hem.* 45.

Rio Janeiro.

26. LOPUS FALLAX.

fallax, *Sgnt. A. S. E. F. 4me Sér.* iii. 570.

Chili.

Eastern Isles.

27. LOPUS PARTILUS.

Niger, nitens ; caput breve ; oculi rufi, prominuli ; rostrum coxas posticas attingens ; antennæ corpore breviores, articulo 1o *rufo,* 2o *dilatato basi rufo ; thorax rufus, nigro antice bimaculatus ; pedes rufi, genubus tarsisque nigris ; corium basi rufum.*

Black, elliptical, smooth, shining. Head short-triangular. Eyes red, prominent. Rostrum extending to the hind coxæ. Antennæ shorter than the body ; first joint red, longer than the head ; second much longer than the first, dilated for more than half the length from the tip, red towards the base; third a little shorter than the first; fourth shorter than the third. Prothorax red, with a large transverse black spot on the fore border. Scutellum red. Legs red; knees and tarsi black. Corium red at the base. Length of the body 3 lines.

a. New Guinea. Presented by W. W. Saunders, Esq.

Australia.

28. Lopus Australis.

Mas et fœm. *Pallide flavus, fusiformis, subtilissime punctatus; caput breve, subsulcatum; oculi prominuli; rostrum coxas posticas attingens; antennæ rufæ, corpore paullo longiores, articulo 2o nigro dilatato basi rufo, 3o basi flavo; prothorax rufo bivittatus; pedes longiusculi; alæ anticæ rufo venosæ, costa nigra, membrana cinerea.*

Male and female. Pale yellow, fusiform, very finely punctured. Head smooth, shining, triangular, with a slight longitudinal furrow between the eyes. Eyes black, prominent. Rostrum extending to the hind coxæ; tip piceous. Antennæ slender, a little longer than the body; first joint red, stout, a little longer than the head; second about thrice the length of the first, red at the base, black and dilated towards the tip; third red, slightly curved, more than half the length of the second, yellow at the base; fourth red, a little more than half the length of the third. Prothorax with a narrow red stripe along each side, which is slightly reflexed. Legs slender, rather long. Corium with red veins and with a black costal stripe, which does not extend to the tip. Membrane and hind wings cinereous. *Male.—* Disk of the abdomen above black. Length of the body 3½—4½ lines.

a—c. Australia. From Mr. Damel's collection.
d. New South Wales. Presented by W. W. Saunders, Esq.
e. New South Wales. From Dr. Stephenson's collection.

29. Lopus sordidus.

Fœm. *Ferrugineus, snbtilissime punctatus; oculi prominuli; rostrum coxas posticas attingens; antennæ corpore paullo breviores, articulo 2o apicem versus nigro dilatato, 3o basi albido; membrana cinerea.*

Female. Ferruginous, elliptical, very finely punctured. Head triangular. Eyes piceous, prominent. Rostrum extending to the hind coxæ; tip black. Antennæ slender, a little shorter than the body; first joint stout, a little shorter than the head; second about twice the length of the first, black and dilated towards the tip; third longer than the first, whitish at the base; fourth as long as the first ventral segments, retracted towards the base of the abdomen. Legs moderately long; femora rather stout Membrane and hind wings cinereous. Length of the body 3 lines.

a. Australia. Presented by the Entomological Club.

Genus 3. PHYTOCORIS.

Phytocoris, *Fall. Hem. Suec.* i. 83. *H.-Sch. Kirschb. Caps.* 30. *Fieb. Crit. Gen.* 26; *Eur. Hem.* 65, 258. *Dougl. and Scott, Hem.* 300.

Europe.
Div. 1.
1. PHYTOCORIS USTULATUS.

Phytocoris nstulatus, *H.-Sch. Nom.* 47. *Fieb. Eur. Hem.* 258.
Bohemia.

2. PHYTOCORIS SIGNORETI.

Signoreti, *Muls. Ann. Soc. Linn.* 1857, 163. *Fieb. Eur. Hem.* 258—
meridionalis? *H.-Sch. Nom.* 48.

France.

3. PHYTOCORIS ALBOFASCIATUS.

albofasciatus, *Fieb. Eur. Hem.* 259.

Switzerland.

4. PHYTOCORIS ULMI.

Cimex Ulmi, *Linn. Syst. Nat.* 503; *Faun. Suec.* 964—Miris Ulmi, *Fabr. Syst. Rhyn.* 256—Phytocoris Ulmi, *Fall. Hem. Suec.* i. 89. *H.-Sch. Wanz. Ins.* iii. 9, pl. 76, f. 234. *Meyer, Caps.* 43, 2. *Kirschb. Caps.* 40. *Flor, Rhyn. Liv.* i. 416; ii. 593. *Fieb. Eur. Hem.* 259. *Dougl. and Scott. Hem.* 313—Var. Phytocoris exoletus, *Costa, Cent.* 1852.

a—n. England. From Mr. Stephens' collection.
o—r. England.
s. Europe.
t. Italy. Presented by F. Walker, Esq.

5. PHYTOCORIS FLAVALIS.

Cimex flavalis, *Fabr. Mant. Ins.* ii. 303—Lygæus flavalis, *Fabr. Ent. Syst.* iv. 171; *Syst. Rhyn.* 235—Lygæus vividus, *Fabr. Syst. Rhyn.* 237—Phytocoris divergens, *Mey. Ent. Zeit. Stett.* 1841, 87; *Caps.* 44, pl. 1, f. 1. *Kirsch. Caps.* 39, 108. *Flor, Rhyn. Liv.* i. 415; ii. 594. *Dougl. and Scott, Hem.* 311. *Fieb. Eur. Hem.* 259—Phytocoris Ulmi, *Fall, Hem. Suec.* i. 89—Miris longicornis, *Wolff, Icon. Cim.* 155, 15, f. 149—Phytocoris Ulmi, *H.-Sch. Nom.* 47—Phytocoris longicornis, *Burm. Handb. Ent.* ii. 269.

Europe.

6. PHYTOCORIS FEMORALIS.

irroratus, *Fieb. Crit.* 3—femoralis, *Fieb. Eur. Hem.* 260.

Corsica.

7. PHYTOCORIS POPULI.

Cimex Populi, *Linn. Faun. Suec.* 963—Lygæus Populi, *Fabr. Syst. Rhyn.*
237—Phytocoris Populi, *Fall. Hem. Suec.* 84. *Kirschb. Caps.* 38.
Fieb. Eur. Hem. 260.
Europe.

8. PHYTOCORIS CRASSIPES.

crassipes, *Flor, Rhyn. Liv.* ii. 606. *Dougl. and Scott, Hem.* 309.
a—n. England. From Mr. Stephens' collection.
o—t. England.
u, v. Europe. From Mr. Children's collection.

9. PHYTOCORIS DIMIDIATUS.

dimidiatus, *Kirschb. Caps.* 122. *Fieb. Eur. Hem.* 260. *Dougl. and Scott,*
Hem. 307—longipennis, *Flor, Rhyn. Liv.* ii. 601.
Europe.

10. PHYTOCORIS TILIÆ.

Lygæus Tiliæ, *Fabr. Syst. Rhyn.* 237—Phytocoris Tiliæ, *Fall. Hem. Suec.*
i. 85. *Mey. Rhyn.* pl. 7, f. 4. *Kirschb. Caps.* 39. *Flor, Rhyn. Liv.*
ii. 599. *Fieb. Eur. Hem.* 260. *Dougl. and Scott, Hem.* 303—Phytocoris Populi, *Meyer, Caps.* pl. 7, 1.
a—d. England. From Mr. Stephens' collection.
e, f. England.
g. Europe. From Mr. Children's collection.
h. Nova Scotia. From Lieut. Redman's collection.
i. St. Martin's Falls. Presented by Dr. Barnston.

11. PHYTOCORIS PINI.

Pini, *Kirschb. Caps.* 22. *Fieb. Eur. Hem.* 261.
Germany.

12. PHYTOCORIS MINOR.

minor, *Kirschb. Rhyn.* 22. *Fieb. Eur. Hem.* 261.
Germany.

13. PHYTOCORIS JUNIPERI.

Juniperi, *Frey-Gessner, Mitth. Schw. Ent. Ges.* 1865, 302.
Switzerland.

14. PHYTOCORIS INCANUS.

incanus, *Fieb. Wien. Ent. Mon.* viii. 326.
Sarepta.

15. PHYTOCORIS DISTINCTUS.

distinctus, *Dougl. and Scott, Hem.* 302.

England.

16. PHYTOCORIS DUBIUS.

dubius, *Dougl. and Scott, Hem.* 305.

England.

17. PHYTOCORIS MARMORATUS.

marmoratus, *Dougl. and Scott, Ent. M. Mag.* v. 261.

England.

18. PHYTOCORIS NOWICKI.

Nowicki, *Fieb. Verh. Zool. Bot. Ges. Wien.* xx. 261.

Galicia.

Div. 2.

Allœonotus, *Fieb. Crit. Gen.* 28, pl. 6, f. 189; *Eur. Hem.* 65, 261.

19. PHYTOCORIS DISTINGUENDUS.

Capsus distinguendus, *H.-Sch. Wanz. Ins.* iv. 33, pl. 121, f. 384—
　　Allœonotus distinguendus, *Fieb. Eur. Hem.* 262.

South Germany.

20. PHYTOCORIS EGREGIUS.

Allœonotus egregius, *Fieb. Wien. Ent. Mon.* viii. 328.

South-East Europe.

Div. 3.

Hallodopus, *Fieb. Crit. Gen.* 29; *Eur. Hem.* 66—Allodapus, *Fieb. Eur·
　　Hem.* 262—Eroticoris, *Dougl. and Scott, Hem.* 471.

21. PHYTOCORIS CORYZOIDES.

Capsus coryzoides, *H.-Sch. Wanz. Ins.* iv. 35, pl. 121, f. 387—Halticus
　　rufescens, *Burm. Handb. Ent.* ii. 278—Capsus brachypterus, *Boh. K.
　　V. Ak. Handl.* 1849, 254—Allodapus coryzoides, *Fieb. Eur. Hem.*
　　262.

North Europe.

Div. 4.

Perideris, *Fieb. Verh. Zool. Bot. Ges. Wien.* xx. 248, pl. 5, f. 6.

22. PHYTOCORIS MARGINATUS.

Perideris margiuata, *Fieb. Verh. Zool. Bot. Ges. Wien.* xx. 249.
Greece.

Note.—According to Fieber, Phytocoris albida, *Kol. Mel.* sp. 109, is
an uncertain species, and does not belong to the genus.

America.

23. PHYTOCORIS MARMORATUS (bis lectum).

Phytocoris marmoratus, *Fitch, MSS.*
a. New York.   Presented by Dr. Asa Fitch.

24. PHYTOCORIS SUBVITTATUS.

subvittatus, *Stal, Rio Jan. Hem.* 47.
Rio Janeiro.

25. PHYTOCORIS EFFICTUS.

effictus, *Stal, Rio Jan. Hem.* 48.
Rio Janeiro.

26. PHYTOCORIS ADSPERSUS.

adspersus, *Spin. Faun. Chil.* 194 ; *Sgnt. A. S. E. F. 4me Sér.* iii. 567—
   marmoratus, *Blanch. Faun. Chil.* 194.
Chili.

27. PHYTOCORIS RUFULUS.

rufulus, *Blanch. Faun. Chil.* 192.   *Sgnt. A. S. E. F. 4me Sér.* iii. 568.
Chili.

28. PHYTOCORIS PALLIDULUS.

pallidulus, *Blanch. Faun. Chil.* 193.   *Sgnt. A. S. E. F. 4me Sér.* iii. 568.
Chili.

29. PHYTOCORIS RUBRESCENS.

rubrescens, *Blanch. Faun. Chil.* 191.   *Sgnt. A. S. E. F. 4me Sér.* iii. 568.
Chili.

30. PHYTOCORIS OBSCURELLUS.

obscurellus, *Blanch. Faun. Chil.* 192.   *Sgnt. A. S. E. F. 4me Sér.* iii.
   569.
Chili.

31. PHYTOCORIS OBSOLETUS.

obsoletus, *Blanch. Faun. Chil.* 194.   *Sgnt. A. S. E. F. 4me Sér.* iii. 569.
Chili.

32. PHYTOCORIS IRRORATUS.

irroratus, *Blanch. Faun. Chil.* 193. *Sgnt. A. S. E. F. 4me Sér.* iii. 569.
Chili.

33. PHYTOCORIS TRIGONALIS.

trigonalis, *Spin. Faun. Chil.* 197. *Sgnt. A. S. E. F. 4me Sér.* iii. 569.
Chili.

34. PHYTOCORIS? COCCINEUS.

P.? coccineus, *Spin. Faun. Chil.* 185, pl. 2, f. 10. *Sgnt. A. S. E. F. 4me Sér.* iii. 570.
Chili.

Genus 4. CYLLOCORIS.

Cyllococoris, *Hahn, Wanz. Ins.* ii. 97. *Kirschb. Caps.* 31. *Fieb. Crit. Gen.* 48; *Eur. Hem.* 69, 282. *Dougl. and Scott, Hem.* 367.

Europe and Siberia.

Div. 1.

1. CYLLOCORIS HISTRIONICUS.

Cimex histrionicus, *Linn. Syst. Nat.* 728. *Sehr. Ins. Austr.* 286—Lygæus agilis, *Fabr. Ent. Syst.* iv. 182. *Wolff, Icon. Cim.* 150, pl. 15, f. 147 —Capsus agilis, *Fabr. Syst. Rhyn.* 247. *Fall, Hem. Suec.* i. 120— Cyllocoris agilis, *Hahn, Wanz. Ins.* ii. 98, pl. 10, f. 182—Phytocoris histrionicus, *Burm. Handb. Ent.* ii. 267—Capsus histrionicus, *Mey. Caps.* 90, 75. *Sahlb. Geoc. Fen.* 96. *Kirschb. Caps.* 43. *Flor, Rhyn. Liv.* i. 475—Polymerus (Kelidocorys) histrionicus, *Kol. Mel. Ent.* ii. 102—Cyllocoris histrionicus, *Fieb. Eur. Hem.* 282. *Dougl. and Scott, Hem.* 368.
Europe.

2. CYLLOCORIS EQUESTRIS.

equestris, *Stall, Stett. Ent. Zeit.* xix. 182.
Siberia.

Div. 2.

Globiceps, *Latr. MS. Serv. Hem. Gen.* 235. *Fieb. Crit. Gen.* 49; *Eur. Hem.* 69. 282. *Dougl. and Scott, Hem.* 362.

3. CYLLOCORIS SPHEGIFORMIS.

Cimex sphegiformis, *Rossi, Faun. Etr. sp.* 1345—Globiceps capito, *Serv. Enc. Meth.* x. 326; *Hist. Hem.* 282, pl. 6, f. 1—Capsus decoratus, *Mey. Caps.* 88. *Kirschb. Caps.* 46—Globiceps sphegiformis, *Fieb. Eur. Hem.* 283.
Europe.

4. CYLLOCORIS PICTETI.

Globiceps Picteti, *Mey. Dür.* *Fieb. Eur. Hem.* 283.
South Spain.

5. CYLLOCORIS DISPAR.

Cyllecoris dispar, *Boh. Nya Sv.* iv. 20—Globiceps dispar, *Fieb. Eur. Hem.* 283.
Sweden.

6. CYLLOCORIS FLAVONOTATUS.

Cyllecoris flavonotatus, *Boh. Nya Sv.* iv. 19—Capsus flavonotatus, *Kirschb. Rhyn.* sp. 32—Lygæus flavomaculatus, *Wolff, Icon. Cim.* 114, pl. 11, f. 108—Capsus flavomaculatus? *Panz. Faun. Germ.* 92, 16—Cyllocoris flavomaculatus, *H.-Sch. Wanz. Ins.* iii. 10, pl. 76, f. 235—Cyllecoris flavonotatus, *Boh. Vet. Ak. Forh.* 1852, 71—Capsus flavonotatus, *Kirschb. Caps.* 47, 109. *Flor, Rhyn. Liv.* i. 467—Globiceps flavonotatus, *Fieb. Crit. Gen.* 49, sp. 13; *Eur. Hem.* 283. *Dougl. and Scott, Hem.* 366.
Europe.

7. CYLLOCORIS FLAVOMACULATUS.

Cimex fulvipes, *Scop. Ent. Carn.* 364—Capsus flavomaculatus, *Panz. Faun. Germ.* 92, 16. *Fabr. Syst. Rhyn.* 247. *Fall. Hem. Suec.* i. 120. *Zett. Ins. Lapp.* 278 *Mey. Caps.* 91, 76. *Sahlb. Geoc. Ins.* 96, 10. *Kirschb. Caps.* 46. *Flor, Rhyn. Liv.* i. 469—Phytocoris flavomaculatus, *Burm. Handb. Ent.* ii. 267—Polymerus (Kelidocoris) flavomaculatus, *Kol. Mel. Ent.* ii. 103—Globiceps flavomaculatus, *Fieb. Crit. Gen.* 49, sp. 13; *Eur. Hem.* 284. *Dougl. and Scott, Hem.* 364.

a. Europe. From Mr. Children's collection.
b, c. Europe.

8. CYLLOCORIS SELECTUS.

selectus, *Fieb. Crit. Gen.* 49, sp. 13; *Eur. Hem.* 284. *Dougl. and Scott, Hem.* 363—*Var.* flavomaculatus?
Germany.

9. CYLLOCORIS INFUSCATUS.

Globiceps infuscatus, *Garb. Bull. Soc. Ent. Ital.* i. 190.
Turin.

10. CYLLOCORIS ATER.

Globiceps ater, *Dougl. and Scott. Ent. M. Mag.* ii. 248.
England.

Div. 3.

Aëtorhinus, *Fieb. Crit. Gen.* 52, pl. 6, f. 8, 31; *Eur. Hem.* 70, 285; *Dougl. and Scott, Hem.* 346.

11. CYLLOCORIS ANGULATUS.

Phytocoris angulatus, *Fall. Hem. Suec.* i. 81. *Zett. Ins. Lapp.* 272—Capsus angulatus, *H.-Sch. Wanz. Ins.* iii. 75, pl. 97, f. 292. *Mey. Caps,* 89, 72. *Sahlb. Geoc. Fen.* 97, 12. *Kirschb. Caps.* 42. *Flor, Rhyn. Liv.* i. 477 — Polymerus (Blephoridopterus) angulatus, *Kol. Mel. Ent.* ii. 108—Aëtorhinus angulatus, *Fieb. Eur. Hem.* 285. *Dougl. and Scott, Hem.* 347.

Europe.

Div. 4.

Plagiorhamma, *Fieb. Verh. Zool. Bot. Ges. Wien.* xx. 250, pl. 6, f. 8.

12. CYLLOCORIS SUTURALIS.

Plagiorhamma suturalis, *Fieb. Verh. Zool. Bot. Ges. Wien.* xx. 250—Capsus suturalis, *H.-Sch. Wanz. Ins.* iv. 32, pl. 120, f. 383.

Hungary.

Div. 5.

Brachyceræa, *Fieb. Crit. Gen.* 93; *Eur. Hem.* 77, 324—Idolocoris, *Dougl. and Scott, Hem.* 374.

13. CYLLOCORIS HYALINIPENNIS.

Phytocoris hyalinipennis, *Klug, Burm. Handb. Ent.* ii. 258—Brachyceræa hyalinipennis, *Fieb. Eur. Hem.* 325.

Portugal. Spain.

14. CYLLOCORIS PALLICORNIS.

Brachyceræa pallicornis, *Pict. Mey. Fieb. Eur. Hem.* 324—Idolocoris pallicornis, *Dougl. and Scott, Hem.* 375.

Spain.

15. CYLLOCORIS ANNULATUS.

Gerris annulatus, *Wolff, Icon. Cim.* 162, pl. 16, f. 156—Capsus annulatus, *H.-Sch. Wanz. Ins.* iii. 52, pl. 88, f. 270. *Mey. Caps.* 58. *Kirschb. Caps.* 47—Brachyceræa annulata, *Fieb. Eur. Hem.* 325—Idolocoris annulatus, *Dougl. and Scott, Hem.* 376.

Europe.

16. Cyllocoris globulifer.

Capsus globulifer, *Fall. Hem. Suec.* i. 124. *Zett. Ins. Lapp.* 377. *Flor, Rhyn. Liv.* i. 512—Capsus alienus, *H.-Sch. Wanz. Ins.* iii. 53, pl. 88, f. 271. *Kirschb. Caps.* 48. *Mey. Caps.* 57—Capsus cyllocoroides, *Scholtz. Arb. Ver.* 133—Brachyceræa globulifera, *Fieb. Eur. Hem.* 325—Idolocoris globulifer, *Dougl. and Scott, Hem.* 377.
Europe.

17. Cyllocoris geniculatus.

Brachyceræa geniculata, *Fieb. Crit. Gen.* 93, Sp. 43; *Eur. Hem.* 325.
Corsica.

Div. 6.

Dicyphus, *Fieb. Crit. Gen.* 94; *Eur. Hem.* 77, 325—Idolocoris, p., *Dougl. and Scott, Hem.*

18. Cyllocoris errans.

Gerris errans, *Wolff, Icon. Cim.* 161, pl. 16, f. 155—Capsus collaris, *Fall. Hem. Suec.* i. 125. *Zett. Ins. Lapp.* 279. *Mey. Caps.* 83, 63. *Kirschb. Caps.* 42, 24. *Flor. Rhyn. Liv.* i. 483—Cyllocoris collaris, *Hahn, Wanz. Ins.* ii. 121, pl. 66, f. 203—Polymerus (Blepharidopterus) collaris, *Kol. Mel. Ent.* ii. 107—Dicyphus errans, *Fieb. Eur. Hem.* 326—Idolocoris collaris, *Dougl. and Scott, Hem.* 379.
Europe.

19. Cyllocoris pallidus.

Capsus collaris, var., *Fall. Hem. Suec.* i. 125—Capsus pallidus, *H.-Sch. Wanz. Ins.* iii. 51, pl. 88, f. 269. *Mey. Caps.* 64. *Kirschb. Caps.* 42—Capsus constrictus, *Boh. K. V. Akad. Forh.* 1852, Sp. 32—Dicyphus pallidus, *Fieb. Eur. Hem.* 326—Idolocoris pallidus, *Dougl. and Scott, Hem.* 380.
Europe.

Div. 7.

Systellonotus, *Fieb. Crit. Gen.* 92, pl. 6, f. 29; *Eur. Hem.* 77, 323. *Dougl. and Scott, Hem.* 369.

20. Cyllocoris triguttatus.

Cimex triguttatus, *Linn. Syst. Nat.* 2729—Lygæus triguttatus, *Fabr. Syst. Rhyn.* 230—Capsus triguttatus, *Fall. Hem. Suec.* i. 121. *Mey. Caps.* 90, 74. *Sahlb. Geoc. Fen.* 92. *Kirschb. Caps.* 51, 110. *Flor. Rhyn. Liv.* i. 480—Cyllocoris triguttatus, *Hahn, Wanz. Ins.* ii. 99, pl. 60, f. 183—Systellonotus triguttatus, *Fieb. Eur. Hem.* 324. *Dougl. and Scott, Hem.* 370.
Europe.

21. CYLLOCORIS THYMI.

Systellonotus Thymi, *Sgnt. A. S. E. F. 4me Sér.* v. 125.
Bourray.

North America.
22. CYLLOCORIS CALIFORNICUS.

Capsus Californicus, *Stal, Eug. Resa,* 259.
California.

Div. Garganus, *Stal, Stett. Ent. Zeit.* xxiii. 321.

23. CYLLOCORIS FUSIFORMIS.

Capsus fusiformis, *Say, Works ed. Leconte,* i. 344—Garganus fusiformis,
  Stal.
United States.

Mexico.
24. CYLLOCORIS ALBIDIVITTIS.

Garganus albidivittis, *Stal, Stett. Ent. Zeit.* xxiii. 322.
Mexico.

South America.

A. Abdomen sessile.
  a. First joint of the antennæ a little shorter than the head
     and the prothorax together.   -   -   -   gracilentus.
  b. First joint of the antennæ as long as the prothorax. -   Amyoti.
  c. First joint of the antennæ nearly as long as the pro-
     thorax. -   -   -   -   -   -   sanguiniceps.
  d. First joint of the antennæ as long as the breadth of the
     head.
   * Coxæ not whitish.
   † Corium spotted. -   -   -   -   -   quadristillatus.
  †† Corium not spotted. -   -   -   -   Costæ.
  ** Coxæ whitish. -   -   -   -   -   bisbistillatus.
  e. First joint of the antennæ shorter than the breadth of
     the head. -   -   -   -   -   stellatipennis.
B. Abdomen almost petiolated. -   -   -   -   petiolatus.

25. CYLLOCORIS GRACILENTUS.

gracilentus, *Stal, Rio Jan. Hem.* 53—Garganus gracilentus, *Stal, Stett.
  Ent. Zeit.* xxiii. 322.
Rio Janeiro.

26. CYLLOCORIS QUADRISTILLATUS.
quædristillatus, *Stal*, *Rio Jan. Hem.* 54.
Rio Janeiro.

27. CYLLOCORIS BISBISTILLATUS.
bisbistillatus, *Stal*, *Rio Jan. Hem.* 54.
Rio Janeiro.

28. CYLLOCORIS STILLATIPENNIS.
stillatipennis, *Stal*, *Ric Jan. Hem.* 54.
Rio Janeiro.

29. CYLLOCORIS SANGUINICEPS.
sanguiniceps, *Stal*, *Rio Jan. Hem.* 54.
Rio Janeiro.

30. CYLLOCORIS COSTÆ.
Costæ, *Stal*, *Rio Jan. Hem.* 54.
Rio Janeiro.

31. CYLLOCORIS AMYOTI.
Amyoti, *Stal*, *Rio Jan. Hem.* 55.
Rio Janeiro.

32. CYLLOCORIS PETIOLATUS.
petiolatus, *Stal*, *Rio Jan. Hem.* 55.
Rio Janeiro.

33. CYLLOCORIS SCUTELLATUS.
Phytocoris scutellatus, *Spin. Ess. Hem.*—Cyllocoris scutellatus, *Sgnt. A. S. E. F. 4me Sér.* iii. 586—Cyllocoris jucundus, *Sgnt. A. S. E. F. 4me Sér.* iii. 570, pl. 11, f. 5.
Chili.

34. CYLLOCORIS CUCURBITACEUS.
Phytocoris cucurbitaceus, *Spin. Faun. Chil.* 195—Cyllocoris cucurbitaceus *Sgnt. A. S. E. F. 4me Sér.* iii. 571.
Chili.

35. CYLLOCORIS LACTEUS.
Phytocoris lacteus, *Spin. Faun. Chil.* 195—Cyllocoris lacteus, *Sgnt. A. S. E. F. 4me Sér.* iii. 571.
Chili.

36. CYLLOCORIS FASCIOLARIS.

Phytocoris fasciolaris, *Blanch. Faun. Chil.* 191—Globiceps fasciolaris,
   Sgnt. A. S. E. F. 4me Sér. iii. 573.
Chili.

Genus 5. CAPSUS.

Capsus, *Fabr. Syst. Rhyn.* 241.   *Kirsch. Caps.* 30.

Europe, Siberia, West Asia.
Div. 1.

Oncognathus, *Fieb. Crit. Gen.* 15; *Eur. Hem.* 63, 246.

1. CAPSUS BINOTATUS.

Capsus binotatus, *Fabr. Syst. Rhyn.* 235.   *H.-Sch. Wanz. Ins.* iii. 77, pl.
   98, f. 296.   *Kirschb. Caps.* 59—Oncognathus binotatus, *Fieb. Eur.
   Hem.* 247.
a. Europe.   From Mr. Children's collection.
b. North Wales.   Presented by F. Walker, Esq.
c. Italy.   Presented by F. Walker, Esq.

Div. 2.

Alloeotomus, *Fieb. Crit. Gen.* 17, pl. 6, f. 23; *Eur. Hem.* 247.

2. CAPSUS GOTHICUS.

Phytocoris gothicus, *Fall. Hem. Suec.* 110—Capsus marginepunctatus,
   H.-Sch. Wanz. Ins. iii. 69, pl. 96, f. 284.   *Kirsch. Rhyn.* 53—Alloeo-
   tomus gothicus, *Fieb. Eur. Hem.* 243.
Europe.

Div. 3.

Pachypterna, *Fieb. Crit. Gen.* 18; *Eur. Hem.* 63, 247.

3. CAPSUS FIEBERI.

Pachypterna Fieberi, *Schmidt, Fieb. Crit. Nov.* 1.   *Eur. Hem.* 248.
Carinthian Alps.

Div. 4.

Bothynotus, *Fieb. Wien. Ent. Mon.* viii. 76, pl. 2, f. 7.

4. CAPSUS MINKI.
Bothynotus Minki, *Fieb. Wien. Ent. Mon.* viii. 77.
Corfu.

Div. 5.
Camptobrochys, *Fieb. Crit. Gen.* 19, pl. 6, f. 4, 35; *Eur. Hem.* 64, 248·
Dougl. and Scott, Hem. 447.

5. CAPSUS FALLENII.
Platycoris Falleni, *Hahn, Wanz. Ins.* ii. 89, pl. 57, f. 175—Phytocoris
punctulatus, *Fall. Mon. Cim.* 87; *Hem. Suec.* i. 95. *Mey. Caps.* 103,
pl. 4, f. 2—Capsus punctulatus, *Sahlb. Geoc.* 112. *H.-Sch. Nom.
Ent.* i. 52. *Kirschb. Caps.* 67. *Flor. Rhyn. Liv.* i. 532—Capsus
Fallenii, *Kirschb. Caps.* 67—Camptobrochys Falleni, *Fieb. Eur.
Hem.* 248.
Europe.

6. CAPSUS PUNCTULATUS.
Phytocoris punctulatus, *Fall. Mon. Cim.* 87; *Hem. Suec.* i. 95. *Mey.
Rhyn.* 103, pl. 4, f. 2—Phytocoris lutescens, *Schill. Verh. Schles.
Ges.* 1836—C. varipennis, *Hoffm.*—Capsus punctulatus, *H.-Sch. Nom.
Ent.* i. 52. *Kirschb. Caps.* 67. *Flor. Rhyn. Liv.* i. 532—Campto-
brochys punctulatus, *Fieb. Eur. Hem.* 249. *Dougl. and Scott, Hem.*
448.
Europe.

7. CAPSUS SERENUS.
Camptobrochys serenus, *Dougl. and Scott, Ent. M. Mag.* v. 135.
Baalbec.

Div. 6.
Stethoconus, *Fieb. Wien. Ent. Mon.* viii. 79, pl. 2, f. 8.

8. CAPSUS MAMILLOSUS.
Capsus cyrtopeltis, *Flor. Rhyn. Liv.* i. 628—Capsus mamillosus, *Flor.
Rhyn. Liv.* ii. 614—Stethocomus mamillosus, *Fieb. Wien. Ent. Mon.*
viii. 79.
Livonia.

Div. 7.
Exœretus, *Fieb. Wien. Ent. Mon.* viii. 81, pl. 2, f. 9.

9. CAPSUS MEYERI.

Camptotylus Meyeri, *Frey, Schw. Ent. Ges.* 1863, 119—Exœretus Meyeri,
    Fieb. Wien. Ent. Mon. viii. 81.

Sarepta.

Div. 8.

Megacœlum, *Fieb. Crit. Gen.* 21; *Eur. Hem.* 64, 249.

10. CAPSUS INFUSUS.

Capsus infusus, *H.-Sch. Wanz. Ins.* iv. 30, pl. 120, f. 381—Phytocoris
    validicornis, *Boh. Vet. Ak. Forh.* 1852, 14, 19—Capsus iufusus,
    Kirschb. Caps. 55—Megacœlum infusum, *Fieb. Eur. Hem.* 249—
    Deræocoris infusus, *Dougl. aud Scott. Hem.* 331.

Europe.

11. CAPSUS AMŒNUS.

Deræocoris amœnus, *Dovgl. and Scott, Ent. M. Mag.* v. 115.

Jordan region.

Div. 9.

Grypocoris, *Dougl. and Scott, Ent. M. Mag.* v. 116.

12. CAPSUS FIEBERI.

Grypocoris Fieberi, *Dougl. and Scott, Ent. M. Mag.* v. 117.

Jordan region.

Div. 10.

Homodemus, *Fieb. Crit. Gen.* 22 ; *Eur. Hem.* 64, 249.

13. CAPSUS FERRUGATUS.

Cimex roseomaculatus, *Deg. Ins.* iii. 193. *Schff. Icon. Rat.* pl. 13, f. 9—
    Cimex digrammus, *Gmel. Linn. Syst. Nat.* 2181—Cimex Ribis et
    rosatus, *Schr. Faun. Boic.* 1149—Cimex cruentatus, *Vill. Linn. Ent.*
    533—Lygæus ferrugatus, *Fabr. Ent. Syst.* iv. 173 ; *Syst. Rhyn.* 236—
    Phytocoris ferrugatus, *Fall. Hem. Suec.* i. 86.   *Hahn, Wanz. Ins.* i.
    204, pl. 33, f. 104.   *Burm. Handb. Ent.* ii. 270.   *Kol. Mel. Ent.* ii.
    111—Capsus ferrugatus, *Sahlb. Geoc. Fen.* 104.   *Meyer, Caps.* 52, 12.
    Kirschb. Caps. 57.   *Flor. Rhyn. Liv.* i. 496—Homodemus ferrugatus,
    Ficb. Eur. Hem. 250—Deræocoris ferrugatus, *Dougl. and Scott, Hem.*
    327.

a—c. North Wales.   Presented by F. Walker, Esq.
d. Isle *of* Wight.   Presented by F. Walker, Esq.
e. Killarney.   Presented by F. Walker, Esq.
f. Polish Ukraine.   Presented by Dr. Dowler.
g. South France.   Presented by F. Walker, Esq.

15. Capsus marginellus.

————, *Sturm, Verz.* pl. 4, f. 5—Miris marginellus, *Fabr. Syst. Rhyn.* 255—Capsus marginellus, *Kirschb. Caps.* 50—Phytocoris scriptus, *Hahn, Wanz. Ins.* ii. 120, pl. 66, f. 202—Homodemus marginellus, *Fieb. Eur. Hem.* 250—Deræocoris marginellus, *Dougl. and Scott, Hem.* 328.

a. England. From Mr. Stephens' collection.
b. England. Presented by F. Walker, Esq.
c—f. England.
g. Geneva.
h. Europe. Presented by W. W. Saunders, Esq.
i—m. Italy. Presented by F. Walker, Esq.

16. Capsus Meyeri.

Lophyrus Meyeri, *Kol. Mel.* 105, pl. 11, f. 26—Homodemus Meyeri, *Fieb. Eur. Hem.* 250.

Caucasus.

17. Capsus angularis.

Homodemus angularis, *Fieb. Wien. Ent. Mon.* viii. 325.

Amasia. Mehadia.

Div. 11.

Brachycoleus, *Fieb. Crit. Gen.* 23, pl. 6, f. 5. *Eur. Hem.* 65, 250.

18. Capsus bimaculatus.

Phytocoris bimaculatus, *Ramb. Faun. And.* 160—Brachycoleus bimaculatus, *Fieb. Eur. Hem.* 251.

Spain.

19. Capsus scriptus.

Lygæus scriptus, *Fabr. Syst. Rhyn.* 234—Capsus scriptus, *Hahn, Wanz. Ins.* iii. 76, pl. 97, f. 294. *Kirsch. Caps.* 59—Brachycoleus scriptus, *Fieb. Eur. Hem.* 251.

a. Chamouni. Presented by F. Walker, Esq.

Div. 12.

Calocoris, *Fieb. Crit. Gen.* 24; *Eur. Hem.* 65, 251.

20. Capsus striatellus.

Lygæus striatellus, *Fabr. Syst. Rhyn.* 236. *Panz. Faun. Germ.* 93, 17—Miris striatellus, *Wolff, Icon. Cim.* 156, pl. 15, f. 150—Phytocoris striatellus, *Fall. Hem. Suec.* i. 84. *Hahn, Wanz. Ins.* ii. 133, pl. 71,

f. 218. *Zett. Ins. Lapp.* 272.   *Kol. Mel. Ent.* ii. 113—Capsus
striatellus, *Mey. Caps.* 94,81.   *Sahlb. Geoc. Fen.* 105.   *Kirsch. Caps.*
56.   *Flor. Rhyn. Liv.* i. 492—Calocoris striatellus, *Fieb. Eur. Hem.*
251—Deræocoris striatellus, *Dougl. and Scott, Hem.* 318.

a—j. England.   From Mr. Stephens' collection.
k. England.   From the collection of the Entomological Society.
l, m. Summit of Snowdon.   Presented by F. Walker, Esq.
n, o. England.
p—s. Europe.   From Mr. Children's collection.

21. CAPSUS BIMACULATUS.

Capsus bimaculatus, *Hoffm. H.-Sch. Nom.* 51 ; *Wanz. Ins.* vi. 48, pl. 196,
f. 607—Phytocoris Schmidti, *Fieb. Weit. B.* i. 102, pl. 2, f. 1—Calocoris bimaculatus, *Fieb. Eur. Hem.* 253.

South Europe.

22. CAPSUS PILICORNIS. '

Capsus pilicornis, *Panz. Faun. Germ.* 99, 22—Capsus anticus, *Muls. Ann.
Soc. Linn.* 1852, 116—Calocoris pilicornis, *Fieb. Eur. Hem.* 252.

Europe.

23. CAPSUS SEXGUTTATUS.

Cimex sexguttatus, *Fabr. Mant. Ins.* ii. 304—Lygæus sexguttatus, *Fabr.
Ent. Syst.* iv. 174; *Syst. Rhyn.* i. 237—Phytocoris sexguttatus, *Fall.
Hem. Suec.* i. 86—Capsus sexguttatus, *H.-Sch. Wanz. Ins.* iii. 77,
pl. 97, f. 295.   *Meyer, Caps.* 92.   *Flor, Rhyn. Liv.* i. 494—Polymerus (Lophyrus) sexguttatus, *Kol. Mel. Ent.* ii. 106—Calocoris sexguttatus, *Fieb. Eur. Hem.* 252—Deræocoris sexguttatus, *Dougl. and
Scott, Hem.* 322.

a—d. Tunis.   From Mr. Fraser's collection.
e, f. Italy.   Presented by Dr. Dowler.
g. Crete.   Presented by W. W. Saunders, Esq.

24. CAPSUS FULVOMACULATUS.

Cimex fulvomaculatus, *Deg. Ins.* iii. 294—Lygæus saltatorius, *Fabr. Syst.
Rhyn.* 239—Phytocoris fulvomaculatus, *Fall. Hem. Suec.* i. 88.   *Zett.
Ins. Lapp.* 273.   *Kol. Mel. Ent.* ii. 199—Capsus fulvomaculatus,
H.-Sch. Wanz. Ins. iii. 50, pl. 87, f. 267; 81, pl. 99, f. 382.   *Meyer,
Caps.* 96.   *Sahlb. Geoo. Fen* 109.   *Flor. Rhyn. Liv.* i. 505—Capsus
fulvomaculatus, *Kirschb. Caps.* 49—Calocoris fulvomaculatus, *Fieb.
Eur. Hem.* 253—Deræocoris fulvomaculatus, *Dougl. and Scott, Hem.*
316.

Siberia.
a—d. Europe.   From Mr. Children's collection.

25. CAPSUS SEXPUNCTATUS.

Lygæus sexpunctatus, *Fabr. Syst. Rhyn.* 224—Lygæus nemoralis, *Fabr. Syst. Rhyn.* 234—Phytocoris Carceli, *Lep. Serv.* 325—Phytocoris sexpunctatus, *Hahn, Wanz. Ins.* i. 131, pl. 69, f. 213, 214, 215—Miris infuscatus? *Brullé, Exp. Mor. Hem.* 77—Calocoris sexpunctatus, *Fieb. Eur. Hem.* 253.

a, b. Geneva. Presented by M. Gonreau.
c—e. South France.
f. Crete. Presented by W. W, Saunders, Esq.
g. Europe. Presented by W. W. Saunders, Esq.
h. Italy. Presented by F. Walker, Esq.
i—l. Tunis. From Mr. Fraser's collection.

26. CAPSUS ALPESTRIS.

Capsus alpestris, *Mey. Rhyn.* 49—Calocoris alpestris, *Fieb. Eur. Hem.* 253 —Deræocoris alpestris, *Dougl. and Scott, Ent. M. Mag.* iv. 47.

England. Swiss and Carinthian Alps.

27. CAPSUS AFFINIS.

Capsus affinis, *H.-Sch. Nom.* 49. *Kirschb. Caps.* 50—Capsus pabulinus, *Mey. Rhyn.* 48, pl. 1, f. 3. *Sahlb. Geoo. Fenn.* 101—Lygæus pabulinus, *Hahn, Wanz. Ins.* i. 148, pl. 23, f. 74—Phytocoris Salviæ, *Hahn, Wanz. Ins.* ii. 133, pl. 61, f. 217—Calocoris affinis, *Fieb. Eur. Hem.* 254.

Europe.

28. CAPSUS VENUSTUS.

Calocoris venustus, *Pict. Mey. Fieb. Eur. Hem.* 254.

Spain.

29. CAPSUS BIPUNCTATUS.

Cimex pabulinus, *Scop. Ent. Carn.* 132 — Lygæus bipunctatus, *Fabr. Syst. Rhyn.* 235—Phytocoris bipunctatus, *Fall. Hem. Suec.* i. 78. *Burm. Handb. Ent.* ii. 270. *Zett. Ins. Lapp.* 271—Capsus bipunctatus, *H.-Sch. Wanz. Ins.* iii. 79, pl. 98, f. 298. *Meyer, Caps.* 51. *Sahlb. Geoc. Fenn.* 101, 20. *Flor, Rhyn. Liv.* i. 498—Capsus bipunctatus, *Kirschb. Caps.* 60—Calocoris bipunctatus, *Fieb. Eur. Hem.* 254—Deræocoris bipunctatus, *Dougl. and Scott, Hem.* 319.

a, b. North Wales. Presented by F. Walker, Esq.
c, d. Paris. From Mr. Children's collection.
e, f. Italy. Presented by F. Walker, Esq.

30. CAPSUS TRIVIALIS.

Capsus trivialis, *Costa, Cent.* 3, 4—Calocoris trivialis, *Fieb. Eur. Hem.* 255.

Italy. Corsica.

31. Capsus Chenopodii.

Miris lævigatus, *Panz. Faun. Germ.* 93, 21. *Wolff, Icon. Cim.* 36, pl. 4,
f. 36—Phytocoris Chenopodii, *Fall. Hem. Suec.* i. 77. *Kol. Mel. Ent.*
ii. 113—Capsus Chenopodii, *Sahlb. Geoe. Fenn.* 100. *Meyer, Caps.*
51, 11. *Kirschb. Caps.* 57. *Flor, Rhyn. Liv.* i. 501—Calocoris
Chenopodii, *Fieb. Eur. Hem.* 255 —Deræocoris Chenopodii, *Dougl.*
and Scott, Hem. 325—Phytocoris binotatus, *Hahn, Wanz. Ins.* i. 202,
pl. 33, f. 103—Capsus brevicollis, *Meyer, Rhyn.* 47, pl. 1, f. 4.

a—c. Europe. From Mr. Children's collection.
d—i. Italy. Presented by F. Walker, Esq.

32. Capsus instabilis.

Calocoris instabilis, *Pict. Mey. Fieb. Eur. Hem.* 255.

Malaga, Spain.

33. Capsus quadripunctatus.

Lygæus quadripunctatus, *Fabr. Syst. Rhyn.* 235—Phytocoris Chenopodii,
var. 4-punctatus, *Fall. Hem. Suec.* 77—Calocoris quadripunctatus,
Fieb. Eur. Hem. 256.

Germany.

34. Capsus Vandalicus.

Cimex Vandalicus, *Rossi, Faun. Etr.* 1743, pl. 7, f. 12—Lygæus Fraxini,
Fabr. Ent. Syst. iv. 172; *Syst. Rhyn.* 236—Capsus Fraxini, *Hahn,*
Wanz. Ins. iii. 82, pl. 99, f. 303—Phytocoris tænioma, *Costa, Cent.*
1852—Capsus Humuli, *Schumml. Arb. Ver.* 1846, 22—Calocoris
Vandalicus, *Fieb. Eur. Hem.* 256.

Europe.

35. Capsus detritus.

Calocoris detritus, *Mey. Fieb. Eur. Hem.* 257.
Switzerland.

36. Capsus Reicheli.

Phytocoris Reicheli, *Fieb. Weit. Beit.* i. 103, pl. 2, f. 2—Calocoris
Reicheli, *Fieb. Crit. Gen.* 2; *Eur. Hem.* 257.

Bohemia. Corinthia.

37. Capsus seticornis.

Lygæus seticornis, *Fabr. Ent. Syst.* iv. 179—Capsus seticornis, *Fabr. Syst.*
Rhyn. 244—Miris seticornis, *Wolff, Icon. Cim.* f. 114—Phytocoris
apicalis, *Hahn, Wanz. Ins.* i. 220, pl. 35, f. 114—Phytocoris lateralis,
Fall. Hem. Suec. i. 88—Miris tibialis, *Wolff, Icon. Cim.* f. 111—
Capsus lateralis, *Kirschb. Caps.* 58—Calocoris seticornis, *Fieb. Eur.*
Hem. 257.

Europe.

38. Capsus Hedenborgi.
Calocoris Hedenborgi, *Fieb. Verh. Zool. Bot. Ges. Wien.* xx. 258.
Bosphorus.

39. Capsus collaris.
Calocoris collaris, *Fieb. Verh. Zool. Bot. Ges. Wien.* xx. 239.
Rhodes. Corfu.

40. Capsus Beckeri.
Calocoris Beckeri, *Fieb. Verh. Zool. Bot. Ges. Wien.* xx. 259.
Sarepta.

41. Capsus Lethierryi.
Calocoris Lethierryi, *Fieb. Verh. Zool. Bot. Ges. Wien.* xx. 260.
France.

42. Capsus nebulosus.
Calocoris nebulosus, *Fieb. Wien. Ent. Mon.* viii. 326.
Lussin.

43. Capsus Kolenatii.
Calocoris Kolenatii, *Fieb. Wien. Ent. Mon.* viii. 219.
Moravia.

44. Capsus fornicatus.
Calocoris fornicatus, *Fieb. Wien. Ent. Mon.* viii. 218—Deræocoris fornicatus,
Dougl. and Scott, Hem. 329.
England.

45. Capsus tetraphlyctis.
Calocoris tetraphlyctis, *Garb. Bull. Soc. Ent. Ital.* i. 184.
North Italy.

46. Capsus rubricosus.
Calocoris rubricosus, *Garb. Bull. Soc. Ent. Ital.* i. 184.
North Italy.

47. Capsus distinguendus.
Calocoris distinguendus, *Garb. Bull. Soc. Ent. Ital.* i. 184.
North Italy.

48. Capsus rubidus.
Calocoris rubidus, *Garb. Bull. Soc. Ent. Ital.* i. 185.
North Italy.

49. CAPSUS ATERRIMUS.

Calocoris aterrimus, *Garb. Bull. Soc. Ent. Ital.* i. 185.
North Italy.

50. CAPSUS ANNULICORNIS.

annulicornis, *Sahlb. Geoc. Fenn.* 100.   *Fieb. Eur. Hem.* 390.
North Europe. . Siberia.

51. CAPSUS AMŒNUS.

Deræocoris amœnus, *Dougl. and Scott, Ent. M. Mag.* 115.
Jordan Region.

Div. 13.

Pycnopterna, *Fieb. Crit. Gen.* 30; *Eur. Hem.* 66. 262.

52. CAPSUS STRIATUS.

Cimex striatus, *Linn. Syst. Nat.* 730; *Faun. Suec.* 960. *Scop. Ent. Carn.*
133. *Deg. Ins.* iii. 191, pl. 15, f. 14, 15. *Schr. Ins. Austr.* 284—
Miris striatus, *Fabr. Syst. Rhyn.* 255. *Wolff, Icon. Cim.* 37, pl. 4. f.
37—Capsus striatus, *Panz. Faun. Germ.* 93; 22. *Meyer, Caps.* 94,
80. *Sahlb. Geoo. Fenn.* 97. *Kirschb. Caps.* 49—Phytocoris striatus,
Fall. Hem. Suec. i. 83. . *Hahn, Wanz. Ins.* ii. 134, pl. 71, f. 219.
Burm. Handb. Ent. ii. 267. *Serv. Hem.* 279. *Zett. Ins. Lapp.* 272.
Flor, Rhyn. Liv. i. 490—Polymerus (Cyllocoris) striatus, *Kol. Mel.*
Ent. ii. 103—Pycnopterna striata, *Fieb. Eur. Hem.* 262—Deræocoris
striatus, *Dougl. and Scott, Hem.* 320.

a—h. England.   From Mr. Stephens' collection.
i. Summit of Snowdon.   Presented by F. Walker, Esq.
j—m. Europe.   From Mr. Children's collection.

53. CAPSUS PULCHER.

Capsus pulcher, *H.-Sch. Wanz. Ins.* iii. 75, pl. 97, f. 293—Pycnopterna
pulchra, *Fieb. Eur. Hem.* 263.
Germany.

Div. 14.

Closterotomus, *Fieb. Crit. Gen.* 27; *Eur. Hem.* 65, 261.

54. CAPSUS BIFASCIATUS.

Lygæus bifasciatus, *Fabr. Ent. Syst.* iv. 177—Capsus bifasciatus, *Fabr.*
Syst. Rhyn. 242. *Sahlb. Geoc. Fenn.* 121. *Kirschb. Caps.* 48—
Phytocoris bifasciatus, *Hahn, Wanz. Ins.* iii. 7, pl. 75, f. 232—
Globiceps variegatus, *Costa*—Capsus Schillingi, *Schumml. Scholz*
Arb. Ver. 1846, 182—Closterotomus bifasciatus, *Fieb. Eur. Hem.* 261.
Europe.

Div. 15.

Gryllocoris, *Bærensp. Berl. Ent. Zeit.* 1859, 334. *Fieb. Eur. Hem.* 66, 263.

55. Capsus angusticollis.

Gryllocoris angusticollis, *Bærensp. Berl. Ent. Zeit.* 1859, 335, pl. 6, f. 8. *Fieb. Eur. Hem.* 263.

Greece.

Div. 16.

Rhopalotomus, *Fieb. Crit. Gen.* 31, pl. 6, f. 38; *Eur. Hem.* 66, 263.

56. Capsus ater.

Cimex ater, *Linn. Syst. Nat.* 725; *Faun. Suec.* 944—Cimex semiflavus, *Linn. Syst. Nat.* 725; *Faun. Suec.* 944—Cimex tyrannus, *Fabr. Sp. Ins.* ii. 371; *Mant. Ins.* ii. 305—Cimex flavicollis, *Fabr. Mant. Ins.* ii. 305—Lygæus ater, *Fabr. Ent. Syst.* iv. 177—Capsus ater, *Fabr. Syst. Rhyn.* 241. *Fall. Hem. Suec.* i. 116. *Hahn, Wanz. Ins.* i. 126, pl. 20, f. 65. *H.-Sch. Nom. Ent.* i. 52. *Burm. Handb. Ent.* 275. *Zett. Ins. Lapp.* 277. *Sahlb. Geoo. Fenn.* 121, 67. *Meyer, Caps.* 108. *Serv. Hem.* 28, i. 486. *Kirschb. Caps.* 54—Capsus tyrannus, *Fabr. Syst. Rhyn.* 242—Lygæus tyrannus, *Wolff, Icon. Cim.* f. 146—Lygæus flavicollis, *Fabr. Ent. Syst.* iv. 178. *Wolff, Icon. Cim.* 32, pl. 4, f. 32 —Capsus flavicollis, *Fabr. Syst. Rhyn.* 243—Heterotoma ater, *Kol. Mel. Ent.* ii. 127—Rhopalotomus ater, *Fieb. Eur. Hem.* 264. *Dougl. and Scott, Hem.* 440.

a—p. England. From Mr. Stephens' collection.
q. England. From Mr. Vigors' collection.
r—v. Europe. From Mr. Children's collection.
w. Europe. Presented by W. W. Saunders, Esq.
x—cc. St. Martin's Falls, Albany River, Hudson's Bay. Presented by Dr. Barnston.
dd, ee. Nova Scotia. From Lieut. Redman's collection.

57. Capsus cinctus.

Capsus cinctus, *Kol. Mel. Ent.* 128, pl. 11, f. 29—Rhopalotomus cinctus, *Fieb. Eur. Hem.* 264.

South Russia.

Div. 17.

Capsus, *Fieb. Crit. Gen.* 32; *Eur. Hem.* 66, 264.

58. Capsus cordiger.

Phytocoris cordiger, *Hahn, Wanz. Ins.* ii. 85, pl. 56, f. 771—Capsus cordiger, *Fieb. Eur. Hem.* 264.

Europe.

59. Capsus punctum.

Phytocoris punctum, *Ramb. Faun. And.* 164—Capsus punctum, *Fieb. Eur. Hem.* 265.

South Spain.

60. Capsus rutilus.

Capsus rutilus, *H.-Sch. Wanz. Ins.* iv. 34, pl. 121, f. 385. *Fieb. Eur. Hem.* 265.

Roumelia, Syria.

61. Capsus Schach.

Cimex Schach, *Fabr. Sp. Ins.* ii. 371; *Mant. Ins.* ii. 305—Lygæus Schach, *Fabr. Ent. Syst.* iv. 177—Capsus Schach, *Fabr. Syst. Rhyn.* 242—Capsus miniatus, *H.-Sch. Wanz. Ins.* iv. 34, pl. 121, f. 381. *Fieb. Eur. Hem.* 265.

a. South France. Presented by the Entomological Club.
b—d. South Europe.
e—j. Italy. Presented by F. Walker, Esq.
k—n. Louvain. Presented by Lady Seymour.   .

62. Capsus trifasciatus.

Cimex trifasciatus, *Linn. Syst. Nat.* 1, 725. *Schff. Icon. Ratisb.* pl. 13, f. 8—Capsus bifasciatus, *Fabr. Syst. Rhyn.* 244. *Kirsch. Caps.* 53. *Fieb. Eur. Hem.* 265—Lygæus elatus, *Fabr. Ent. Syst.* iv. 176. *Wolff, Icon. Cim.* f. 31. *Panz. Faun. Germ.* 73, 20—Capsus elatus, *Fabr. Syst. Rhyn.* 241.

Europe.

63. Capsus annulipes.

annulipes, *H.-Sch. Wanz. Ins.* vi. 97, pl. 212, f. 669. *Fieb. Eur. Hem.* 265.

South Europe.

64. Capsus olivaceus.

Capsus olivaceus, *Fabr. Syst. Rhyn.* 214. *Fieb. Eur. Hem.* 266—Capsus medius, *Kirsch. Caps.* 52—Capsus rufipes, *Fabr. Syst. Rhyn.* 242.

Europe.

65. Capsus cardinalis.

cardinalis, *Fieb. Crit. Sp.* 4; *Eur. Hem.* 266.

Bohemia.

66. Capsus scutellaris.

Lygæus scutellaris, *Fabr. Ent. Syst.* iv. 180—Capsus scutellaris, *Fabr. Syst. Rhyn.* 245. *Coq. Ill. Icon.* pl. 19, f. 8. *Flor, Rhyn. Liv.* i. 510. *Fieb. Eur. Hem.* 266. *Dougl. and Scott, Hem.* 443. Phytocoris scutellaris, *Zett. Act. Holm.* 1819. *Fall. Hem. Suec.* i. 109.

Europe.

67. Capsus capillaris.

Cimex capillaris, *Fabr. Mant. Ins.* ii. 305—Cimex tricolor, *Fabr. Mant. Ins.* ii. 306—Lygæus capillaris, *Fabr. Ent. Syst.* iv. 180—Lygæus Danicus, *Fabr. Ent. Syst.* iv. 181. *Wolff. Icon. Cim.* 34, pl. 4, f. 34 —Lygæus tricolor, *Fabr. Ent. Syst.* iv. 181. *Wolff. Icon. Cim.* 35, pl. 4, f. 35—Capsus capillaris, *Fabr. Syst. Rhyn.* 244. *Burm. Handb. Ent.* ii. 274. *Serv. Hist. Hem.* 281. *Fieb. Eur. Hem.* 266. *Dougl. and Scott, Hem.* 442—Capsus Danicus, *Fabr. Syst. Rhyn.* 246. *Hahn, Wanz. Ins.* i. 17, pl. 2, f. 9—Capsus tricolor, *Fabr. Syst. Rhyn.* 246. *Panz. Faun. Germ.* 93, 20. *H.-Sch. Nom. Ent.* i. 51. *Meyer, Caps.* 108, 98. *Kirschb. Caps.* 52. *Flor, Rhyn. Liv.* i. 509—Phytocoris Danicus, *Fall. Hem. Suec.* i. 109.

a—l. England. From Mr. Stephens' collection.
m. France.
n—p. South France. Presented by F. Walker, Esq.
q. Geneva. Presented by M. Goureau.
r. Polish Ukraine. Presented by Dr. Dowler.

68. Capsus coruscus.

coruscus, *Garbiglietti, Bull. Soc. Ent. Ital.* i. 186.

Sardinia.

Div. 18.

Dioneus, *Fieb. Eur. Hem.* 67, 268.

69. Capsus neglectus.

Capsus neglectus, *Fabr. Syst. Rhyn.* 242. *H.-Sch. Wans. Ins.* iii. 82, pl. 100, f. 304—Dioneus neglectus, *Fieb. Eur. Hem.* 269.

South Europe.

70. Capsus infuscatus.

Miris infuscatus, *Brullé, Exp. Mor.* 77—Dioneus infuscatus, *Fieb. Eur. Hem.* 269.

Greece.

Div. 19.

Camptoneura, *Fieb. Crit. Gen.* 35—Campyloneura, *Fieb. Eur. Hem.* 67, 269. *Dougl. and Scott, Hem.* 372.

71. CAPSUS VIRGULA:

Capsus virgula, *H.-Sch. Wanz. Ins.* iii. 51, pl. 88, f. 268—Campyloneurá virgula, *Fieb. Eur. Hem.* 269. *Dougl. and Scott, Hem.* 373.

Europe.

Div. 20.

Dichroscytus, *Fieb. Crit. Gen.* 36. *Eur. Hem.* 67, 269. *Dougl. and Scott,* Hem. 477—Dichroscytidæ, *Dougl. and Scott, Hem.*

72. CAPSUS RUFIPENNIS.

Phytocoris rufipennis, *Foll. Hem. Suec.* i. 92. *Zett. Ins. Lapp.* 274— Capsus rufipennis, *H.-Sch. Wanz. Ins.* vi. 50, pl. 197, f. 610. *Sahlb. Geoo. Fenn.* 105. *Kirschb. Caps,* 551. *Flor, Rhyn. Liv.* i. 489— Dichroscytus rufipennis, *Fieb. Eur. Hem.* 270. *Dougl. and Scott,* Hem. 478.

a. England. Presented by F. Walker, Esq.
b. Europe.

73. CAPSUS VALESIANUS.

Capsus Valesianus, *Meyer, Catal.*—Dichroscytus Valesianus, *Fieb. Eur. Hem.* 270.

Vallais.

Div. 21.

Liocoris, *Fieb. Crit. Gen.* 37, pl. 6, f. 15 ; *Eur. Hem.* 67, 270. *Dougl. and Scott, Hem.* 449.

74. CAPSUS TRIPUSTULATUS.

Cimex tripustulatus, *Fabr. Sp. Ins.* ii. 370; *Mant. Ins.* ii. 304—Lygæus tripustulatus, *Fieb. Ent. Syst.* iv. 176; *Syst. Rhyn.* 239—Phytocoris tripustulatus, *Fall. Hem. Suec.* i. 96. *Hahn, Wanz. Ins.* i. 215, pl. 34, f. 111. *Burm. Handb. Ent.* ii. 273. *Zett. Ins. Lapp.* 275. *Kol. Mel. Ent.* ii. 120—Capsus tripustulatus, *H.-Sch. Nom. Ent.* i. 52. *Meyer, Caps.* 106. *Sahlb. Geoc. Fenn.* 113. *Kirschb. Caps.* 64, 112. *Flor, Rhyn. Liv.* i. 515—Phytocoris Pastinacæ, *Hahn, Wanz. Ins.* 213, pl. 34, f. 110—Liocoris tripustulatus, *Fieb. Eur. Hem.* 271. *Dougl. and Scott, Hem.* 450.

a—e. England. From Mr. Stephens' collection.
f. England. Presented by F. Walker, Esq.

Div. 22.

Polymerus, *Hahn, Wanz. Ins.* i. 27. *Fieb. Crit. Gen.* 39; *Eur. Hem.* 67, 271—Systratiotus, *Dougl. and Scott, Hem.* 442.

75. CAPSUS INTERMEDIUS.

Caps ᵘ intermedius, *Sahlb. Geoc. Fenn.* 116—Polymerus intermedius, *Fieb Eur. Hem.* 391.

Finland.

76. Capsus holosericeus.

Polymerus holosericeus, *Hahn, Wans. Ins.* i. 27, pl. 4, f. 17. *Fieb. Eur. Hem.* 271—Capsus holosericeus, *Kirschb. Rhyn.* 74.

Europe.

77. Capsus nigritus.

Phytocoris nigritus, *Fall. Hem. Suec.* i. 97—Capsus nigritus, *H.-Sch. Nom. Ent.* i. 52; *Wanz. Ins.* vi. 45, pl. 195, f. 601. *Sahlb. Geoc. Fenn.* 116. *Kirschb. Caps.* 69, 113—Capsus nigrita, *Flor, Rhyn. Liv.* i. 547—Polymerus nigritus, *Fieb. Eur. Hem.* 391—Systratiotes nigritus, *Dougl. and Scott, Hem.* 444.

Europe.

Div. 23.

Charagochilus, *Fieb. Crit. Gen.* 38; *Eur. Hem.* 67. *Dougl. and Scott, Hem.* 445.

78. Capsus Gyllenhalii.

Phytocoris Gyllenhalii, *Fall. Mon. Cim.* 88 ; *Hem. Suec.* i. 97. *Zett. Ins. Lapp.* 225—Capsus Gyllenhalii, *Hahn, Mon. Ent.* i. 56 ; *Wanz. Ins.* iii. 86, pl. 101, f. 310. *Mey. Caps.* 61, 28 ; *Sahlb. Geoc. Fenn.* 116, 57. *Kirschb. Caps.* 69, 114—Charagochilus Gyllenhali, *Fieb. Eur. Hem.* 271—Charagochilus Gyllenhalii, *Dougl. and Scott, Hem.* 446.

Europe. Siberia.

Div. 24.

Cyphodema, *Fieb. Crit. Gen.* 40 ; *Eur. Hem.* 68, 272.

79. Capsus Meyer-Düri.

Cyphodema Meyer-Düri, *Fieb. Crit. Gen. Sp.* 5 ; *Eur. Hem.* 272.
Corsica.

Div. 25.

Tylonotus, *Fieb. Crit. Gen.* 41—Plesiocoris, *Fieb. Eur. Hem.* 272.

80. Capsus rugicollis.

Phytocoris rugicollis, *Fall. Hem. Suec.* 79—Capsus rugicollis, *Hahn, Wanz. Ins.* iii. 80, pl. 98, f. 299. *Kirschb. Nachtrag.* 55--Plesiocoris rugicollis, *Fieb. Eur. Hem.* 272—Phytocoris marginatus, *Boh.*

a. Europe. From Mr. Children's collection.

Div. 26.

Lygus, *Hahn, Wanz. Ins.* i. 147. *Fieb. Crit. Gen.* 42 ; *Eur. Hem.* 68, 272. *Dougl. and Scott, Hem.* 456.

81. Capsus pratensis.

Cimex pratensis, *Linn. Syst. Nat.* 728; *Faun. Suec.* 949. *Scop. Ent. Carn.* 133—Lygæus pratensis, *Fabr. Ent. Syst.* iv. 174; *Syst. Rhyn.* 234. *Fall. Mon. Cim.* 83—Lygæus umbellatorum, *Panz. Faun. Germ.* 93, 19—Phytocoris pratensis, *Fall.·Hem. Suec.* i. 90. *Hahn, Wanz. Ins.* i. 217, pl. 35, f. 112. *Zett. Ins. Lapp.* 273. *Kol. Mel. Ent.* ii. 119, 99—Phytocoris alpina, *Kol. Mel. Ent.* ii. 20, 100, pl. 10, f. 24—Capsus gemellatus, *H.-Sch, Wanz. Ins.* iii. 81, pl. 99, f. 301. *Kirschb. Caps.* 64, 102—Lygus pratensis, *Fieb. Eur. Hem.* 273. *Dougl. and Scott, Hem.* 464—Capsus pratensis, *Sahlb. Geoc. Fenn.* 111. *Kirschb. Caps.* 64, 112. *Flor, Rhyn. Liv.* i. 517.

Europe. Siberia.

82. Capsus campestris.

Cimex campestris, *Linn. Syst. Nat.* 728; *Faun. Suec.* 950—Lygæus campestris, *Fabr. Mant. Ins.* ii. 303; *Ent. Syst.* iv. 171; *Syst. Rhyn.* 234—Capsus Artemisiæ. *Schilling*—Phytocoris campestris, *Fall. Hem. Suec.* i. 91. *Hahn, Wanz. Ins.* i. 218, pl. 35, f. 113. *Zett. Ins. Lapp.* 273. *Kol. Mel. Ent.* ii. 118—Capsus campestris, *Sahlb. Geoc. Fenn.* 111. *Kirschb. Caps.* 65—Capsus pratensis (ex parte), *Flor, Rhyn. Liv.* i. 517—Lygus campestris, *Fieb. Eur. Hem.* 273. *Dougl. and Scott, Hem.* 463.

Europe. Siberia.

83. Capsus rubricatus.

Lygæus rubricatus, *Fall. Mon. Cim.* 91—Phytocoris rubricatus, *Fall. Hem. Suec.* i. 100. *Zett. Ins. Lapp.* 275—Capsus rubricatus, *Sahlb. Geoc. Fenn.* 106. *Kirschb. Caps.* 58, 111. *Flor, Rhyn. Liv.* i. 526—Lygus rufescens, *Hahn, Wanz. Ins.* i. 28, pl. 4, f. 18—Capsus rubicundus, *Mey. Caps.* 72, 44—Lygus rubricatus, *Hahn, Wanz. Ins.* i. 153, pl. 24, f. 80. *Fieb. Eur. Hem.* 274, 392. *Dougl. and Scott, Hem.* 462.

Europe.

84. Capsus atomarius.

Hadrodema atomaria, *Meyer Dür. Caps.*—Capsus delicatus, *Muls. Ann. Soc. Lin.* 1857, 167—Lygus atomarius, *Fieb. Eur. Hem.* 392.

France.

85. Capsus limbatus.

Phytocoris limbatus, *Fall. Hem. Suec.* 92. *Ahr. Faun. Eur.* 13, 20—Lygus limbatus, *Hahn, Wanz. Ins.* i. 182, pl. 23, f. 77. *Fieb. Eur. Hem.* 274—Capsus limbatus, *Kirschb. Nachtrag.* 66—Phytocoris viridis, *Fall. Hem. Suec.* i. 93—Capsus viridis, *Mey. Rhyn.* 77.

Europe. Siberia.

86. Capsus contaminatus.

Lygæus contaminatus, *Fall. Mon. Cim.* 76—Phytocoris contaminatus, *Fall. Hem. Suec.* i. 75—Lygus contaminatus, *Hahn, Wanz. Ins.* i. 151, pl. 23, f. 76. *Fieb. Crit.* 7; *Eur. Hem.* 274. *Dougl. and Scott, Hem.* 461—Capsus contaminatus, *Mey. Caps.* 45. *Flor, Rhyn. Liv.* i. 531—Capsus sulcifrons, *Kirschb. Caps.* 130, 183—Lygus sulcifrons, *Dougl. and Scott, Hem.* 460.

Siberia.

a. Europe. From Mr. Children's collection.

87. Capsus lucorum.

Capsus lucorum, *Meyer Dür. Caps.* 46, pl. 6, f. 2—Lygus contaminatus, *Kirschb. Caps.* 64, 183.

According to Fieber Phytocoris lucorum, *Boh. K. V. Acad. Forh* 1852, 15, is not this species.

Germany.

88. Capsus commutatus.

Lygus commutatus, *Fieb. Grit. Gen.* 72, sp, 7; *Eur. Hem.* 274.

Switzerland.

89. Capsus apicalis.

Lygus apicalis, *Pict. Mey. Fieb. Eur. Hem.* 275.

South Spain.

90. Capsus Spinolæ.

Capsus Spinolæ, *Mey. Stett. Ent. Zeit.* 1841, 86; *Caps.* 45, pl. 1, f. 2—Lygus Spinolæ, *Fieb. Eur. Hem.* 275. *Dougl. and Scott, Hem.* 458.

England. Germany. Switzerland.

91. Capsus lucorum (bis lectum).

Phytocoris lucorum, *Boh. K. V. Acad. Forh.* 1852, 15.

Sweden.

Not lucorum, *Meyer Dür.*

92. Capsus pabulinus.

Cimex pabulinus, *Linn. Syst. Nat.* 727; *Faun. Suec.* 947—Miris pabulinus, *Fabr. Syst. Rhyn.* 254—Phytocoris pabulinus, *Fall. Hem. Suec.* i. 79. *Zett. Ins. Lapp.* 272—Capsus pabulinus, *Sahlb. Geoc. Fenn.* 101. *Kirschb. Cops.* 57. *Flor, Rhyn. Liv.* i. 507—Capsus affinis, *Mey. Caps.* 48, 6, pl. 1, f. 5—Phytocoris affinis, *Kol. Mel.*

Ent. ii. 116—Lygus pabulinus, *Fieb. Eur. Hem.* 276. *Dougl. and Scott, Hem.* 457.

Siberia.

a—f. Isle of Wight. Presented by F. Walker, Esq.
g. North Wales. Presented by F. Walker, Esq.

93. CAPSUS PUTONI.

Lygus Putoni, *Mey. Dür. Mitth. Schweiz. Ent. Ges.* iii. 207.

Switzerland.

94. CAPSUS FLAVOVIRENS.

Lygus flavovirens, *Fieb. Eur. Hem.* 276.

Switzerland.

95. CAPSUS CHLORIS.

Capsus affinis, *Scholz. Arb. Ver.* (1846)—Capsus viridis, *Mey. Cat.*— Lygus chloris, *Fieb. Crit. Gen.* 42, sp. 8 ; *Eur. Hem.* 276.

Germany. Switzerland.

Div. 27.

Pœciloscytus, *Fieb. Crit. Gen.* 43; *Eur. Hem.* 68, 276. *Dougl. and Scott, Hem.* 466.

96. CAPSUS UNIFASCIATUS.

Lygæus unifasciatus, *Fabr. Ent. Syst.* iv. 187—Capsus unifasciatus, *Fabr. Syst. Rhyn.* 243. *H.-Sch. Nom. Ent.* i. 51. *Mey. Caps.* 104, 93. *Sahlh. Geoe. Fenn.* 108. *Flor, Rhyn. Liv.* i. 544. *Kirschb. Caps.* 61—Miris semiflavus, *Wolff, Icon. Cim.* 154, pl. 15, f. 148— Phytocoris semiflavus, *Fall. Hem. Suec.* i. 86. *Hahn, Wanz. Ins.* i. 208, pl. 34, f. 107—Phytocoris lateralis, *Hahn, Wanz. Ins.* ii. 85, pl. 56, f. 169—Phytocoris marginatus, *Hahn, Wanz. Ins.* ii. 85, pl. 56, f. 170—Phytocoris unifasciatus, *Kol. Mel. Ent.* ii. 123—Pœciloscytus unifasciatus, *Fieb. Eur. Hem.* 276. *Dougl. and Scott, Hem.* 467.

Siberia.

a. Europe.

97. CAPSUS VULNERATUS.

Lygæus vulneratus, *Wolff, Panz. Faun. Germ.* (1801) 100, 22—Phytocoris Dalmanni, *Fall. Hem. Suec.* 87. *Hahn, Wanz. Ins.* i. 210, pl. 34, f. 108—Capsus Dalmanni, *Kirschb. Caps.* 63—Pœciloscytus vulneratus, *Fieb. Eur. Hem.* 277.

Siberia.

a. Europe. From Mr. Children's collection.
b. ———— ?

98. Capsus cognatus.

Pœciloscytus cognatus, *Fieb. Crit. Gen.* 43, sp. 6 ; *Eur. Hem.* 277.
Hungary.

Div. 28.

Hadrodema, *Fieb. Crit. Gen.* 44 ; *Eur. Hem.* 68, 277.

99. Capsus rubicundus.

Phytocoris rubicundus, *Fall. Hem. Suec.* 92 — Capsus rubicundus, *Kirschb. Caps.* 68—Lygus rufescens, *Hahn, Wanz. Ins.* i. 28, pl. 4, f. 18—Capsus rubicundus, *Meyer, Rhyn.* sp. 45—Hadrodema rubicunda, *Fieb. Eur. Hem.* 278.

Siberia.

a. Europe.

100. Capsus Pinastri.

Phytocoris Pinastri, *Fall. Hem. Suec.* 112—Capsus Pinastri, *Kirschb. Caps.* 54. *Sahlb. Geoc. Fenn.* 118—Hadrodema Pinastri, *Fieb. Eur. Hem.* 278.

Europe.

Div. 29.

Orthops, *Fieb. Crit. Gen.* 45, pl. 6, f. 10 ; *Eur. Hem.* 68, 278. *Dougl. and Scott, Hem.* 451.

101. Capsus montanus.

Capsus montanus, *Schill. Arb. Ver.* (1836). *Scholz. Arb. Ver.* 33—Capsus fasciatus,|*Mey. Stett. Ent. Zeit.* ii. (1841)'86; *Rhyn.* pl. 5, f. 5. *H.-Sch. Wanz. Ins.* vi. 99, pl. 212, f. 671 Orthops montanus, *Fieb. Eur. Hem.* 279.

Germany. Spain.

102. Capsus Foreli.

Capsus Foreli, *Mey. Cat.*—Orthops Foreli, *Fieb. Beit.* sp. 9 ; *Eur. Hem.* 279.

Switzerland.

103. Capsus Pastinacæ.

Lygæus Pastinacæ, *Mon. Cim.* 86—Phytocoris Pastinacæ, *Fall. Hem. Suec.* i. 94—Lygæus transversalis? *Fabr. Syst. Rhyn.* 238—Capsus Pastinacæ, *Sahlb. Geoc. Fenn.* 113. *Flor, Rhyn. Liv.* i. 523—Capsus lucidus, *Kirschb. Caps.* 68, 113—Orthops Pastinacæ, *Fieb. Eur. Hem.* 279. *Dougl. and Scott, Hem.* 455.

Sitka.

a. England.
b. Bellagio. Presented by F. Walker, Esq.
c, d. Europe. From Mr. Children's collection.

104. Capsus pellucidus.

Orthops pellucidus, *Fieb. Crit. Gen.* 45, sp. 10; *Eur. Hem.* 279.
Switzerland.

105. Capsus cervinus.

Capsus cervinus, *H.-Sch. Wanz. Ins.* vi. 57, pl. 197, f. 617. *Mey. Caps.*
103, 91. *Kirschb. Caps.* 62—Orthops cervinus, *Fieb. Eur. Hem.* 279.
Dougl. and Scott, Hem. 454.
England. Germany.

106. Capsus flavovarius.

Lygæus flavovarius, *Fabr. Ent. Syst.* iv. 178. *Fall. Mon. Cim.* 86—
Capsus flavovarius, *Fabr. Syst. Rhyn.* 243—Phytocoris flavovarius,
Fall. Hem. Suec. i. 93. *Hahn, Wanz. Ins.* 120, pl. 34, f. 109. *Burm.*
Handb. Ent. ii. 272—Orthops flavovarius, *Fieb. Eur. Hem.* 280.

a. Europe.

107. Capsus Kalmii.

Cimex Kalmii, *Linn. Syst. Nat.* 728; *Faun. Suec.* 148—Capsus Kalmii,
H.-Sch. Nom. Ent. i. 52. *Mey. Caps.* 105, 95. *Sahlb. Geoc. Fenn.*
112. *Kirschb. Caps.* 66, 113. *Flor, Rhyn. Liv.* i. 521—Capsus
transversalis, *H.-Sch. Nom. Ent.* i. 52—Phytocoris Kalmii, *Zett. Ins.*
Lapp. 274. *Kol. Mel. Ent.* ii. 122—Capsus pauperatus, *H.-Sch.*
Wanz. Ins. iv. 31, pl. 120, f. 382—Orthops Kalmi, *Fieb. Eur. Hem.*
280—Orthops Kalmii, *Dougl. and Scott, Hem.* 452.

a. Europe.

Div. 30.

Pilophorus, *Hahn*—Camaronotus, *Fieb. Crit. Gen.* 79, pl. 6, f. 28; *Eur.*
Hem. 74, 313. *Dougl. and Scott, Hem.* 358.

108. Capsus clavatus.

Cimex clavatus, *Linn. Syst. Nat.* 729—Capsus bifasciatus, *Fabr. Syst.*
Rhyn. 242—Phytocoris bifasciatus, *Hahn, Wanz. Ins.* iii. 7, pl. 75, f.
232. *Fall, Hem. Suec.* i. 118—Phytocoris clavatus, *Burm. Handb.*
Ent. ii. 266—Capsus clavatus, *H.-Sch. Wanz. Ins.* iii. 47. *Zett. Ins.*
Lapp. 278. *Mey. Caps.* 87. 70. *Sahlb. Geoc. Fenn.* 91. *Kirschb.*
Caps. 72, 137. *Flor, Rhyn. Liv.* i. 569—Camaronotus bifasciatus,
Fieb. Eur. Hem. 314. *Dougl. and Scott, Hem.* 360—Phytocoris
sphegiformis, *Kol. Mel. Ent.* ii. 110.

a—c. England. From Mr. Stephens' collection.
d. England. Presented by F. Walker, Esq.
e. Europe. From Mr. Children's collection.

109. Capsus confusus.

Capsus confusus, *Kirschb. Caps.* 72—Capsus clavatus, *H.-Sch. Wanz. Ins.*
iii. 47, pl. 87, f. 264—Camaronotus confusus, *Fieb. Eur. Hem.* 314.
Germany.

110. Capsus cinnamopterus.

Capsus cinnamopterus, *Kirschb. Caps.* 72, 135, 10.   *Flor, Rhyn. Liv.* i.
572—Camaronotus cinnamopterus, *Fieb. Eur. Hem.* 314.   *Dougl.
and Scott, Hem.* 359.

Germany.   Switzerland.

a—d. England.   From Mr. Stephens' collection.
e. England.   Presented by F. Walker, Esq.

Div. 31.

Mimocoris, *Scott, Ent. M. Mag.* viii. 195.

111. Capsus camaranotoides.

Mimocoris camaranotoides, *Scott, Ent. M. Mag.* viii. 195.
Corsica.

Div. 32.

Phylus, *Hahn, Wanz. Ins.* i. 26.   *Fieb. Crit. Gen.* 81 ; *Eur. Hem.* 75,314.
Dougl. and Scott. Hem. 354.

112. Capsus lituratus.

Cyllocoris lituratus, *Eversm. MSS.*—Phylus lituratus, *Fieb. Crit. Gen.* 81,
sp. 33; *Eur. Hem.* 315.

Ural Region.

113. Capsus palliceps.

Phylus palliceps, *Fieb. Eur. Hem.* 315.   *Dougl. and Scott, Hem.* 355.
England.   Spain.

114. Capsus melanocephalus.

Cimex melanocephalus, *Linn. Syst. Nat.* 728—Cimex pallens, *Fabr.
Mant. Ins.* ii. 306—Miris pallens, *Fabr. Ent. Syst.* iv. 185; *Syst.
Rhyn.* 254—Lygus melanocephalus, *Hahn, Wanz. Ins.* i. 155, pl. 24,
f. 79—Phytocoris melanocephalus, *Burm. Handb. Ent.* ii. 268—
Phytocoris revestitus, *Fall. Hem. Suec.* 89—Capsus melanocephalus,
Mey. Caps. 55, 17.   *Sahlb. Geoc. Fenn.* 99.   *Kirschb. Caps.* 74.
Flor. Rhyn. Liv. i. 621—Capsus nigriceps, *Muls. Ann. Soc. Linn.*
1852, 118—Phylus melanocephalus, *Fieb. Eur. Hem.* 315.   *Dougl.
and Scott, Hem.* 355.

Note.—Phytocoris nigriceps, *Fall. Hem. Suec.* 104, and P. nigriceps,
Boh. K. V. Akad. Forh. 1852, 15, are one species, which is unknown to
Fieber.

a, b. Europe.   From Mr. Children's collection.

115. Capsus Coryli.

Cimex Coryli, *Linn. Syst. Nat.* 733; *Faun. Suec.* 974—Lygæus Coryli, *Fabr. Ent. Syst.* iv. 171; *Syst. Rhyn.* 234—Phytocoris Coryli, *Fall. Hem. Suec.* 190. *Burm. Handb. Ent.* ii. 268—Capsus Coryli, *Mey. Caps.* 54, 15. *Sahlb. Geoc. Fenn.* 99. *Kirschb. Caps.* 74. *Flor, Rhyn. Liv.* i. 620—Phylus pallipes, *Hahn, Wanz. Ins.* i. 26, pl. 4, f. 16—Phylus Coryli, *Fieb. Eur. Hem.* 315. *Dougl. and Scott, Hem.* 356.

a. Europe.

116. Capsus Avellanæ.

Capsus Avellanæ, *H.-Sch. Wanz. Ins.* vi. 98, pl. 212, f. 670. *Mey. Caps.* 16, pl. 2, f. 2. *Kirschb. Caps.* 75—Phylus Avellanæ, *Fieb. Eur. Hem.* 315. *Dougl. and Scott, Hem.* 357.

Europe.

117. Capsus nigricollis.

Phylus nigricollis, *Garb. Bull. Soc. Ent. Ital.* i. 193.

Turin.

Div. 33.

Zygimus, *Fieb. Verh. Zool. Bot. Ges. Wien.* xx. 249, pl. 6. f. 7.

118. Capsus nigriceps.

Phytocoris nigriceps, *Fall. Hem.* 104. *Boh. Kongl. V. Ak. Forh.* 1852, 15 —Zygimus nigriceps, *Fieb.*

Sweden.

Div. 34.

Pithanus, *Fieb. Crit. Gen.* 16, pl. 6, f. 13; *Eur. Hem.* 61, 239. *Dougl. and Scott, Hem.* 280.

119. Capsus Märkeli.

Capsus Märkeli, *H.-Sch. Wanz. Ins.* iv. 78, pl. 132, f. 406. *Kirschb. Rhyn.* 28—Capsus flavolimbatus, *Boh. K. V. Acad. Handl.* 1849, 252 —Cyllocoris vittatus, *Dahlb. K. V. Acad. Handl.* 205, 1850—Capsus Märkelii, *Flor, Rhyn. Liv.* i. 513—Pithanus Märkeli, *Fieb. Eur. Hem.* 239. *Dougl. and Scott, Hem.* 280.

Europe.

120. Capsus Marshalli.

Pithanus Marshalli, *Dougl. and Scott, Ent. M. Mag.* v. 114.

Nazareth.

Div. 35.

Acropelta, *Mella, Bull. Soc. Ent.* i. 292.

121. Capsus Pyri.

Acropelta *Pyri, Mella, Bull. Soc. Ent. Ital.* i. 203, pl. 4.
Lombardy.

Div. ?
122. Capsus suturalis.

Capsus suturalis, *H.-Sch. Wanz. Ins.* iv. 32, pl. 120, f. 383. *Fieb. Eur. Hem.* 391.
Hungary.

123. Capsus triannulatus.

Deræocoris triannulatus, *Stal, Stett. Ent. Zeit.* xix. 183.
Siberia.

124. Capsus nigronasutus.

Deræocoris nigronasutus, *Stal, Stett. Ent. Zeit.* xix. 184.
Siberia.⁻

125. Capsus illotus.

Deræocoris illotus, *Stal, Stett. Ent. Zeit.* xix. 184.
Siberia.

126. Capsus approximatus.

Deræocoris approximatus, *Stal, Stett. Ent. Zeit.* xix. 185.
Sitka.

127. Capsus brachialis.

brachialis. *Stal, Stett. Ent. Zeit.* xix. 185.
Siberia.

128. Capsus mutans.

Deræocoris mutans, *Stal, Stett. Ent. Zeit.* xix. 186.
Siberia.

129. Capsus simulans.

Deræocoris simulans, *Stal, Stett. Ent. Zeit.* xix. 186.
Siberia.   Kamtschatka.

130. Capsus Halimocnemis.

Halimocnemis, *Becker, Bull. Soc. Nat. Mosc.* xxxvii. 485.
Sarepta.

131. Capsus Pyrethri.
Pyrethri, *Becker, Bull. Soc. Nat. Mosc.* xxxvii. 485.
Sarepta.

132. Capsus Freyi.
Freyi, *Becker, Bull. Soc. Nat. Mosc.* xxxvii. 485.
Sarepta.

133. Capsus Artemisiæ.
Artemisiæ, *Becker, Bull. Soc. Nat. Mosc.* xxxvii. 487.
Sarepta.

134. Capsus desertus.
desertus, *Becker, Bull. Soc. Nat. Mosc.* xxxvii. 487.
Sarepta.

135. Capsus Volgensis.
Volgensis, *Becker, Bull. Soc. Nat. Mosc.* xxxvii. 488.
Sarepta.

136. Capsus apicalis.
apicalis, *Sgnt. A. S. E. F.* 4me Sér. v. 125.
South France.

137. Capsus miniatus.
miniatus, *Parfitt, Ent. M. Mag.* ii. 130.
Devonshire.

138. Capsus coruscus.
coruscus, *Garb. Bull. Soc. Ent. Ital.* i. 186.
Sardinia.

North America.

A. Head black.
  a. Head wholly black or blackish.
    * Prothorax *ochraceous.* -   - -   xanthomelas.
    ** Prothorax with two ochraceous ban*ds.* -   -   incisus.
  *** Prothorax black or blackish.
    † Body *broad.* -   -   -   -   caligineus.
    †† Body narrow. -   -   -   -   obscurellus.
  b. Head with yellow stripes.   --   -   -   strigulatus.
  c. Vertex ferruginous. -   -   -   -   marginatus.
B. Head piceous. -   -   -   -   -   filicornis.
C. Head re*d.*
  a. Head wholly re*d.*
    * Membrane black or blackish

  † Scutellum black. - - - -   coccineus.
  †† Scutellum red. - - - - -   Floridanus.
  ** Membrane cinereous, varied with brown. -   pallescens.
  b. Middle lobe of the head black. - - -   frontifer.
  D. Head luteous, yellow or testaceous.
  a. Membrane not marked.
  * Membrane blackish.
  † Head with black lines. - - - -   Robiniæ.
  †† Head with no black lines. - - -   limbatus.
  ** Membrane pale cinereous. - - -   stramineus.
  b. Membrane spotted. - - - -   hirsutulus.
  c. Membrane streaked. - - - -   contiguus.

139. CAPSUS LINEOLARIS.

Coreus? lineolaris, *Pal. Beauv. Ins.* 187, pl. 11, f. 7—(linearis).

a—g. Nova Scotia. From Lieut. Redman's collection.
h. New York. Presented by Dr. Asa Fitch.

140. CAPSUS EXTERNUS.

externus, *H.-Sch. Wanz. Ins.* viii. 16, pl. 254, f. 791.

a. St. John's Bluff, East Florida. Presented by E. Doubleday, Esq.

141. CAPSUS MELAXANTHUS.

melaxanthus, *H.-Sch. Wanz. Ins.* viii. 18, pl. 254, f. 794.

a—d. Nova Scotia. From Lieut. Redman's collection.
e. Nova Scotia. Presented by W. W. Saunders, Esq.
f—h. Trenton Falls, New York. Presented by E. Doubleday, Esq.
i—k. North America. Presented by F. Walker, Esq.
l. Cincinnati. Presented by T. G. Lea, Esq.
m. Erie. Presented by E. Doubleday, Esq.

142. CAPSUS DIVISUS.

divisus, *H.-Sch. Wanz. Ins.* ix. 167, pl. 313, f. 960.

a. St. John's Bluff, East Florida. Presented by E. Doubleday, Esq.

143. CAPSUS RAPIDUS.

rapidus, *Say, Works ed Leconte,* i. 339—multicolor, *H.-Sch. Wanz. Ins.*
  viii. 19, pl. 294, f. 795.

a. Canada. Presented by Dr. Barnston.
b—e. Nova Scotia. From Lieut. Redman's collection.
f, g. New York. Presented by E. Doubleday, Esq.
h. New York. Presented by the Entomological Club.
i. North America. Presented by F. Walker, Esq.
j. Salem. Presented by Dr. Asa Fitch.

144. CAPSUS CROCEIPES.

croceipes, *H.-Sch. Wanz. Ins.* viii. 16, pl. 254, f. 792.

a. Illinois. Presented by the Entomological Club.
b. Illinois. Presented by E. Doubleday, Esq.

145. CAPSUS ROBINIÆ.

Robiniæ, *Uhler, Proc. Ent. Soc. Phil.* i. 24.
Maryland.

146. CAPSUS RUBROVITTATUS.

Restbenia rubrovittata, *Stal, Stett. Ent. Zeit.* xxiii. 318.
North America.

147. CAPSUS XANTHOMELAS.

Mas et fœm. *Ater, fusiformis, subtilissime punctatus, prothorax scutello propectoreque ochraceis; oculi subprominuli; rostrum coxas intermedias attingens; antennæ corpore paullo breviores; prothorax antice bisulcatus; membrana nigricans.*

Male and female. Deep black, fusiform, very finely punctured. Head triangular. Eyes slightly prominent. Rostrum extending to the middle coxæ. Antennæ a little shorter than the body; first joint longer than the head; second more than twice the length of the first; third much shorter than the second; fourth a little shorter than the first. Prothorax and scutellum bright ochraceous, the former with two transverse furrows near the fore border Propectus ochraceous. Legs slender. Membrane and hind wings blackish. Length of the body 3½ lines.

a, b. Lake Huron. From Dr. Bigsby's collection.
c. St. John's Bluff, East Florida. Presented by E. Doubleday, Esq.

148. CAPSUS CALIGINEUS.

caligineus, *Stal, Eug. Resa,* 258.
California.

149. CAPSUS PUNCTIPES.

Phytocoris punctipes, *Fitch, MSS.*
a. New York. Presented by Dr. Asa Fitch.

150. CAPSUS ALBINERVUS.

Phytocoris albinervus, *Fitch, MSS.*
a. New York. Presented by Dr. Asa Fitch.

151. CAPSUS INCISUS.

Fœm. *Ater, ellipticus, subtilissime punctatus; oculi subprominuli; rostrum coxas posticas attingens; antennæ corpori æquilongæ; prothorax fasciis duabus strigaque lanceolata ochraceis; scutellum ochraceum.*

Female. Deep black, elliptical, very finely punctured. Head triangular. Eyes piceous, slightly prominent. Rostrum extending to the hind coxæ. Antennæ as long as the body; first joint longer than the head; second very much longer than the first. Prothorax with two ochraceous bands; first band broad, notched, emitting a lanceolate streak nearly to the second, which is narrow and abbreviated, and on the hind border. Scutellum ochraceous. Legs moderately long and slender. Length of the body 3½ lines.

a. St. John's Bluff, East Florida. Presented by E. Doubleday, Esq.

152. CAPSUS COCCINEUS.

Mas. Coccineus, fusiformis, subtilissime punctatus, scutello pedibus membranaque nigris; oculi subprominuli; rostrum piceum, coxas intermedias attingens; antennæ nigræ, corpore breviores, articulis 1o 2oque subdilutatis; prothorax antice bicallosus.

Male. Bright red, fusiform, very finely punctured. Head triangular. Eyes piceous, slightly prominent. Rostrum piceous, extending to the middle coxæ. Antennæ black, shorter than the body; first and second joints slightly dilated; first as long as the head; second more than twice the length of the first; third a little longer than the first; fourth much shorter than the third. Prothorax with a callus on each side near the fore border. Scutellum, legs, membrane and hind wings black. Legs slender. Length of the body 2¼ lines.

a. St. John's Bluff, East Florida. Presented by E. Doubleday, Esq.

153. CAPSUS LIMBATELLUS.

Fœm: Niger, ellipticus, nitens, subtilissime punctatus, capite prothoracis margine antico lateribusque coriique vitta costali flavis; oculi subprominuli; rostrum flavum, coxas posticas attingens; prothorax transverse subsulcatus; pedes flavi; membrana nigricans.

Female. Black, elliptical, shining, very finely punctured. Head, fore part and sides of the prothorax, legs and costal stripe of the corium yellow. Head triangular. Eyes piceous, slightly prominent. Rostrum yellow, extending to the hind coxæ. Antennæ slender; first joint as long as the head; second about twice the length of the first. Prothorax with a slight transverse furrow. Legs yellow, slender. Membrane blackish. Length of the body 2½ lines.

a. Trenton Falls, New York. Presented by E. Doubleday, Esq.

154. CAPSUS OBSCURELLUS.

Mas. Niger, gracilis, fere linearis, subtilissime punctatus; caput sat magnum; oculi subprominuli; rostrum fulvum; antennæ piceæ; prothorax antice subsulcatus; pedes picei, graciles; corium nigricans; membrana cinerea.

Male. Black, slender, nearly linear, very finely punctured. Head triangular, rather large. Eyes piceous, slightly prominent. Rostrum tawny. Antennæ piceous, slender; first joint a little shorter than the head; second more than twice the length of the first. Prothorax with a slight transverse furrow. Legs piceous, slender. Corium blackish. Membrane cinereous. Length of the body 1¾ line.

a. St. Martin's Falls, Albany River, Hudson's Bay. Presented by Dr. Barnston.

155. Capsus strigulatus.

Fœm. *Niger, ellipticus, nitens, subtilissime punctatus, subtus flavo bivittatus; caput flavo quadrivittatum; oculi subprominuli; rostrum flavum, coxas posticas attingens; antennæ luteæ, corpore breviores; prothorax luteo marginatus univittatus et quadristrigatus; scutellum fuscum, litura flava trifurcata; pedes postici lutei, tibiis basi femoribusque nigris; corium fuscum, flavo strigatum; membrana cinerea, fusco nebulosa.*

Female. Black, elliptical, shining, very finely punctured. Head triangular, with four yellow stripes above. Eyes red, slightly prominent. Rostrum yellow, extending to the hind coxæ. Antennæ luteous, slender, shorter than the body; first joint nearly as long as the head; second about twice the length of the first; third not longer than the first; fourth shorter than the third. Prothorax with a narrow luteous border, and with a slender luteous stripe, on each side of which there are two longitudinal luteous streaks; fore border reflexed. Scutellum brown, with a trifurcate yellow mark. Pectus and under side of the abdomen with a yellow stripe on each side. Hind legs luteous; femora black; tibiæ black towards the base. Corium brown, with some yellow streaks. Membrane cinereous, mottled with brown. Length of the body 2 lines.

a. Canada. Presented by Dr. Barnston.

156. Capsus frontifer.

Fœm. *Rufus, fusiformis, nitens, subtilissime punctatus; capitis lobus intermedius niger; oculi subprominuli; rostrum coxas intermedias attingens; antennæ nigræ, articulo 1o basi rufo; prothorax antice subsulcatus et bicallosus; membrana fuscescens.*

Female. Red, fusiform, shining, very finely punctured. Head triangular; middle lobe black. Eyes black, slightly prominent. Rostrum extending to the middle coxæ; tip black. Antennæ black; first joint a' little longer than the head, red at the base. Prothorax with a slight transverse furrow, in front of which there is a callus on each side. Legs slender. Membrane and hind wings brownish. Length of the body 3½ lines.

a. North America. Presented by the Entomological Club.

157. Capsus pallescens.

Fœm. *Flavus, ellipticus, nitens, subtilissime punctatus; caput rufum; oculi subprominuli; rostrum coxas posticas attingens; antennæ gracillimæ, corpore paullo breviores; prothorax fusco bistrigatus, antice bicallosus; abdomen rufum; femora quatuor anteriora apice rufa; femora postica rufa, basi pallide flava; corium apicem versus fusco suffusum, apice albidum; membrana pallide cinerea, fusco nebulosa.*

Female. Yellow, elliptical, shining, very finely punctured. Head red, triangular. Eyes piceous, slightly prominent. Rostrum extending to the hind coxæ; tip brown. Antennæ very slender, a little shorter than the body; first joint as long as the head; second more than twice the length of the first; third a little longer than the first. Prothorax with a callus on each side near the fore border and with a short contiguous brown streak on each side. Abdomen red. Legs slender; four anterior femora red towards the tips; hind femora red, pale yellow towards the base. Corium diffusedly tinged with brown towards the tip, which is whitish. Membrane pale cinereous, partly clouded with brown. Hind wings pale cinereous. Length of the body 2 lines.

a. St. Martin's Falls, Albany River, Hudson's Bay. Presented by Dr. Barnston.

158. Capsus hirsutulus.

Mas et fœm. *Luteus aut testaceus, fusco conspersus, subtus pallide flavus; oculi subprominuli; rostrum coxas posticas attingens; antennæ corpore vix breviores, articulo 1o hirsuto, 2o apice fusco; prothorax antice bicallosus; pedes hirsutuli, femoribus apice fusco conspersis, tibiis bifasciatis; corium macula subapicali lutea aut albida; membrana fuscescente cinerea, macula costali alba.*

Male and female. Luteous or testaceous, fusiform, minutely and more or less speckled with brown, pale yellow beneath. Eyes piceous, slightly prominent. Rostrum extending to the hind coxæ. Antennæ nearly as long as the body; first joint stout, hirsute; second brown towards the tip, much longer than the first. Prothorax with a callus on each side near the fore border, which is reflexed. Legs slender, slightly hirsute; femora speckled with brown towards the tips; tibiæ with two darker bands, the first basal. Corium with a luteous or whitish costal subapical spot. Membrane brownish cinereous, with a white costal spot. Length of the body 3 lines.

a, b. Lake Huron. From Dr. Bigsby's collection.

159. Capsus contiguus.

Fœm. *Pallide flavus, fusiformis, subtilissime punctatus; oculi subprominuli; rostrum coxas posticas attingens; antennæ corpori æquilongæ; prothorax antice bicallosus; membrana cinerea, fusco pallido strigata.*

Female. Straw-colour, fusiform, very finely punctured. Head triangular. Eyes piceous, slightly prominent. Rostrum extending to the hind coxæ. Antennæ slender, as long as the body; first joint as long as the head; second more than twice the length of the first; third shorter than the first; fourth shorter than the third. Prothorax with a callus on each side near the fore border. Legs slender. Membrane cinereous, with some pale brown streaks. Length of the body 2¼ lines.

a. Trenton Falls, New York. Presented by E. Doubleday, Esq.

160. Capsus stramineus.

Mas et fœm. *Pallide flavus, fusiformis, subtilissime punctatus; oculi prominuli; rostrum coxas posticas attingens; antennæ corpori æquilongæ; prothorax antice subsulcatus; membrana pallidissime cinerea.*

Male and female. Straw-colour, fusiform, very finely punctured. Head triangular. Eyes piceous, prominent. Rostrum extending to the hind coxæ; tip black. Antennæ slender, as long as the body; first joint as long as the head; second a little more than twice the length of the first; third as long as the first; fourth a little shorter than the third. Prothorax with a slight transverse furrow near the fore border. Legs slender. Membrane and hind wings very pale cinereous. Length of the body $2\frac{3}{4}$—3 lines.

a—e. Nova Scotia. From Lieut. Redman's collection.

161. Capsus filicornis.

Fœm. *Piceus, fusiformis, subtiliter punctatus, subtus fulvus; oculi subprominuli; rostrum fulvum, coxas posticas paullo superans; antennæ fulvæ, filiformes, corpori æquilongæ; prothoracis sulcus transversus bene determinatus; pedes fulvi; corium fulvum, apice rufo unimaculatum fusco triguttatum; membrana fuscescens.*

Female. Piceous, fusiform, finely punctured, tawny beneath. Head triangular. Eyes slightly prominent. Rostrum tawny, extending a little beyond the hind coxæ. Antennæ tawny, filiform, as long as the body; first joint much longer than the head; second more than twice the length of the first; third a little longer than the first; fourth much shorter than the third. Prothorax with a strongly-marked transverse furrow. Legs tawny, slender. Corium tawny; tip with a red spot and with three brown dots. Membrane brownish. Length of the body 4 lines.

a. St. John's Bluff, East Florida. Presented by E. Doubleday, Esq.

162. Capsus marginatus.

Fœm. *Niger, ellipticus nitens, subtilissime punctatus; vertex ferrugineus; oculi prominuli; rostrum fulvum, coxas posticus attingens; antennæ corpori æquilongæ; prothoracis latera rufa; tibiæ tarsique fulva; corii costa late rufa; membrana nigricans.*

Female. Black, elliptical, shining, very finely punctured. Head triangular; vertex ferruginous. Eyes piceous, prominent. Rostrum tawny, extending to the hind coxæ. Antennæ as long as the body; first joint a little longer than the head; second very much longer than the first. Prothorax with red sides and with a callus on each side in front. Legs moderately long and slender; tibiæ and tarsi tawny. Corium with a very broad red costal stripe. Membrane blackish. Length of the body $2\frac{1}{2}$ lines.

a. Trenton Falls, New York. Presented by E. Doubleday, Esq.

163. Capsus Floridanus.

Fœm. *Rufus, fusiformis, subtilissime punctatus; oculi prominuli; rostrum coxas posticas attingens; antennæ nigræ, corpori æquilongæ; prothorax antice bicallosus; pedes picei; membrana nigricans.*

Female. Red, fusiform, very finely punctured. Head triangular. Eyes piceous, prominent. Rostrum extending to the hind coxæ. Antennæ black, slender, as long as the body; first joint red, as long as the head; second more than twice the length of the first; third much shorter than the second; fourth less than half the length of the third. Prothorax with a callus on each side in front. Legs piceous, slender. Membrane blackish. Hind wings cinereous. Length of the body 2 lines.

a, b. St. John's Bluff, East Florida. Presented by E. Doubleday, Esq.

Mexico.

Div. Resthenia.

A. Sides of the prothorax, at least in front, slightly attenuated and reflexed. Scutellum slightly convex (Platytylus, *Fieb.*)

a. Prothorax with a black transverse spot. - - plagigera.
b. Prothorax black, with a luteous border. - - luteigera.
c. Prothorax with two black spots and a black hind border. - - - - - picticollis.

B. Sides of the prothorax wholly obtuse, not reflexed before the middle. Scutellum not or slightly convex.

a. Prothorax luteous, black and with a luteous stripe hindward. - - - - - ornaticollis.
b. Prothorax luteous-red, with a black middle band. - Högbergi.
c. Prothorax luteous, with two black stripes which are abbreviated in front. - - - - bivittis.
D. Prothorax red, with a black stripe, which is much widened hindward. - - - - latipennis.
e. Prothorax black, with a whitish hind border. vittifrons, vitticeps.

164. Capsus melanochrus.

melanochrus, *H.-Sch. Wanz. Ins.* viii. 17, pl. 254, f. 793.
Mexico.

165. Capsus tetrastigma.

tetrastigma, *H.-Sch. Wanz. Ins.* ix. 166, pl. 313, f. 959.
Mexico.

a. —— ? From Mr. Children's collection.

166. Capsus picticollis.

Resthenia picticollis, *Stal, Stett. Ent. Zest.* xxiii. 317.
Mexico.

167. Capsus divisus.

divisus, *H.-Sch. Wanz. Ins.* ix. 167, pl. 313, f. 960—Resthenia divisa,
Stal, Stett. Ent. Zeit. xxiii. 317.
Mexico.

168. Capsus plagigera.

Resthenia plagigera, *Stal, Stett. Ent. Zeit.* xxiii. 316.
Mexico.

169. Capsus luteiger.

Resthenia luteigera, *Stal, Stett. Ent. Zeit.* xxiii. 317.
Mexico.

170. Capsus ornaticollis.

Resthenia ornaticollis, *Stal, Stett. Ent. Zeit.* xxiii. 317.
Mexico.

171. Capsus Högbergi.

Resthenia Högbergi, *Stal, Stett. Ent. Zeit.* xxiii. 317.
Mexico.

172. Capsus bivittis.

Resthenia bivittis, *Stal, Stett. Ent. Zeit.* xxiii. 318.
Mexico.

173. Capsus latipennis.

Resthenia latipennis, *Stal, Stett. Ent. Zeit* xxiii. 318.
Mexico.

174. Capsus vittifrons.

Resthenia vittifrons, *Stal, Stett. Ent. Zeit.* xxiii. 318.
Mexico.

175. Capsus vitticeps.

Resthenia vitticeps, *Stal, Stett. Ent. Zeit.* xxiii. 318—vittifrons, fem.?
Mexico.

Mexico.
Div. Brachycoleus, *Fieb.*

A. Prothorax olive-green, yellowish in front.   -   -   alacer.
B. Prothorax yellow, with two black spots. -   -   -   nigriger.
C. Prothorax olive-green, with four black spots.   -   -   ornatulus.

176. Capsus alacer.

Brachycoleus alacer, *Stal, Stett. Ent. Zeit.* xxiii. 319.
Mexico.

177. Capsus nigriger.

Brachycoleus nigriger, *Stal. Stett. Ent. Zeit.* xxiii. 319.
Mexico.

178. Capsus ornatulus.

Brachycoleus ornatulus, *Stal, Stett. Ent. Zeit.* xxiii. 319.
Mexico.

Div. Calocoris.

A. Second joint of the antennæ twice the length of the first.   jurgiosus.
B. Second joint of the antennæ more than twice the length
of the first. -   -   -   -   -   fasciativentris.

179. Capsus jurgiosus.

Calocoris jurgiosus, *Stal, Stett. Ent. Zeit.* xxiii. 320.
Mexico.

180. Capsus fasciativentris.

Calocoris fasciativentris, *Stal, Ent. Zeit.* xxiii. 320.
Mexico.

Div. Megacœlum, *Fieb.*

181. Capsus rubrinervis.

rubrinervis, *Stal, Stett. Ent. Zeit.* xxiii. 321.
Mexico.

Div. Lygus, *Hahn.*

182. Capsus Sallei.

Lygus Sallei, *Stal, Stett. Ent. Zeit.* xxiii. 321.
Mexico.

183. Capsus scitulus.

Mas. *Niger, fusiformis, subtilissime punctatus ; caput apud oculos
luteum ; oculi prominuli ; rostrum coxas posticas attingens ; antennæ
corpore paullo breviores ; prothorax luteus, nigro unifasciatus et
uniplagiatus ; pectus luteum ; corium basi vittaque costali luteis.*

Male. Black, fusiform, very finely punctured. Head short-triangular,
luteous about the eyes and about the sockets of the antennæ. Eyes
piceous, prominent. Rostrum extending to the hind coxæ. Antennæ
slender, a little shorter than the body ; first joint as long as the head ;
second more than twice the length of the first ; third a little longer than

the first. Prothorax luteous, with a transverse furrow and with an anterior black band, which is connected with a black discal patch, the latter extending to the hind border. Pectus luteous. Legs black, slender. Corium luteous at the base and with a luteous costal stripe. Length of the body 2 lines.

a. Oajaca. From M. Sallé's collection.

184. CAPSUS OPACUS.

Mas. *Niger, fusiformis, nitens, subtilissime punctatus; oculi sat prominuli; rostrum coxas posticas attingens; antennæ corpore paullo breviores; prothorax antice bicallosus et transverse sulcatus; corium vitta costali rufa; membrana nigricans.*

Male. Black, fusiform, shining, very finely punctured. Head triangular. Eyes rather prominent. Rostrum extending *to* the hind coxæ. Antennæ a little shorter than the body; first joint as long as *the* head; second more than twice *the* length of the first; third a little shorter than the first. Prothorax with a slight callus on each side near *the* fore border and in front of a slight transverse furrow. Corium with a red costal stripe. Membrane blackish. Length of the body 2½—3 lines.

a. Mexico. Presented by E. P. Coffin, Esq.
b. Mexico. Presented by W. W. Saunders, Esq.

185. CAPSUS DECORATUS.

Luteus, ellipticus, glaber, nitens; oculi prominuli; rostrum coxas anticas paullo superans; prothorax antice bicallosus, postice viridescens et nigro bimaculatus; pectus nigro quadriplagiatum; pedes pallide flavi, coxis ex parte nigris, femoribus posticis nigro unifasciatis; corium nigro unistrigatum, uniplagiatum et uniguttatum; membrana pallide cinerea, basi fusca lineam arcuatum pallide cineream includente.

Luteous, elliptical, smooth, shining. Head triangular. Eyes piceous, prominent. Rostrum extending a little beyond the fore coxæ; tip black. Prothorax greenish towards the hind border, near which there is a large black spot on each side; a callus on each side in front. Pectus with two black patches on each side. Legs pale yellow, moderately long and slender; coxæ partly black; hind femora with a black band beyond the middle. Corium with a black streak extending from the base along the interior border, connected with a black exterior discal patch; a more exterior black discal dot. Membrane pale cinereous; basal part brown, including a transverse curved pale cinereous line. Length of the body 3 lines.

a. Orizaba. From M. Sallé's collection.

186. CAPSUS BICINCTUS.

Mas. *Ater, fusiformis, subtilissime punctatus; oculi prominuli; rostrum coxas intermedias attingens; antennæ corpore multo breviores; prothorax antice ochraceus, transverse bisulcatus; pectus ochraceum, atro biplagiatum; corium basi et scutellum ochracea.*

Male. Deep black, fusiform, very finely punctured. Head triangular. Eyes prominent. Rostrum extending to the middle coxæ. Antennæ much shorter than *the* body; first joint longer than the head; second more than twice the length of the first; third as long as the first. Prothorax with a broad ochraceous band along the fore border, including two strongly-marked transverse furrows and having a notch on its hind border. Pectus ochraceous, with a deep black patch on each side. Corium *at* the base and scutellum ochraceous. Legs slender,'moderately long. Length of the body 4 lines.

a. Oajaca. From M. Sallé's collection.

West Indies.
187. CAPSUS DIMIDIATUS.
dimidiatus, *Guér. Hist. Fis. Cuba*, vii. 168.

Cuba.

Div. n.
Heterocoris, *Guér. Hist. Fis. Cuba*, vii. 164.

188. CAPSUS DILATATUS.
H. dilatata, *Guér. Hist. Fis. Cuba*, vii. 164.

Cuba.

189. CAPSUS JAMAICENSIS.

Fœm. Læte rufus, fusiformis, subtilissime punctatus; capitis lobus intermedius antice niger; oculi subprominuli; rostrum coxas intermedias attingens; antennæ nigræ, graciles, corpore breviores; prothoracis sulcus transversus bene determinatus; pedes longiusculi; alæ atræ.

Female. Bright red, fusiform, very finely punctured. Head triangular; middle lobe black in front. Eyes black, slightly prominent. Rostrum black, extending *to* the middle coxæ. Antennæ black, slender, shorter than the body; first joint a little longer than the head; second more than twice the length of the first; third shorter than the first; fourth shorter than the third. Prothorax with a strongly-marked transverse furrow. Legs black, slender, rather long. Wings black. Length of the body 3 lines.

a. Jamaica. Presented by W. W. Saunders, Esq.
b. Jamaica. From Mr. Gosse's collection.

South America.
Div. 1.

Deræocoris, *Kirschb.*

A. Corium not vitreous.
a. Costa of corium not dilated.
* Membrane not spotted.
† Corium not speckled.

‡ Antennæ not very long.
§ Eyes not very large.
✕ First joint of the antennæ not very short.
o Prothorax blackish or blackish brown.
＋ Prothorax striped. -   -   -   -   -   cribricollis.
＋＋ Prothorax not striped.
＋＋ Head blackish.
ɷ Corium banded. -   -   -   -   -   nobiliatus.
ɷɷ Corium not banded.
＋ Antennæ pale towards the base. -   -   -   testaceipes.
＋＋ Antennæ not pale towards the base.   -   -   luctuosus.
＋＋＋＋ Head pale, middle lobe black.   -   -   -   caligatus.
oo Prothorax pale.
＋ Prothorax grayish white. -   -   -   illotus.
＋＋ Prothorax not grayish white.
＋＋ Hind tibiæ not blackish.
ɷ Corium blackish.
＋ Femora not banded.   -   -   -   -   Wallengreni.
＋＋ Femora banded. -   -   -   -   -   fraudulentus.
ɷɷ Corium pale.
＋ Femora banded.
＝ Scutellum pale. -   -   -   -   -   vittiscutis.
＝＝ Scutellum blackish brown.   -   -   -   cribratus.
＋＋ Femora not banded.
＝ Antennæ not blackish at the base.
V Membrane with brownish veins.
Λ Antennæ black. -   -   -   -   -   viridicans.
ΛΛ Antennæ pale, with black bands. -   -   -   ciucticornis.
VV Membrane with red veins. -   -   - fuscomaculatus.
＝＝＝ Antennæ blackish at the base.   -   -   basicornis.
＋＋＋＋ Hind tibiæ blackish. -   -   -   - semiochraceus.
✕✕ First joint of the antennæ very short.   -   -   cribrosus.
§§ Eyes very large. -   -   -   -   -   sticticollis.
‡‡ Antennæ much longer than the body.   -   -   purgatus.
†† Corium speckled. -   -   -   -   -   lenticulosus.
** Membrane spotted. -   -   -   -   sticticus.
b. Costa of corium dilated.
* Second joint of the antennæ thrice or more than
thrice as long as the first.   -   -   dilatatus.
** Second joint of the antennæ hardly thrice as long
as the first. -   -   -   -   -   Dahlbomi.
*** Second joint of the antennæ twice as long as the
first. -   -   -   -   -   -   fraudans.
B. Corium vitreous.
a. Corium whitish vitreous. -   -   -   vitreus.
b. Corium not whitish.
* Body blackish. -   -   -   -   -   clarus.
** Body whitish yellow. -   -   -   -   insignis.

190. Capsus nobilitatus.
Deræocoris nobilitatus, *Stal, Rio Jan. Hem*. 48.
Rio Janeiro.

191. Capsus Wallengreni.
Deræocoris Wallengreni, *Stal, Rio Jan. Hem*. 48.
Rio Janeiro.

192. Capsus cribricollis.
Deræocoris cribricollis, *Stal, Rio Jan. Hem*. 48.
Rio Janeiro.

193. Capsus vittiscutis.
Deræocoris vittiscutis, *Stal, Rio Jan. Hem*. 48.
Rio Janeiro.

194. Capsus fraudulentus.
Deræocoris fraudulentus, *Stal, Rio Jan. Hem*. 49.
Rio Janeiro.

195. Capsus viridicans.
Deræocoris viridicans, *Stal, Rio Jan. Hem*. 49.
Rio Janeiro.

196. Capsus fuscomaculatus.
Deræocoris fuscomaculatus, *Stal, Rio Jan. Hem*. 49.
Rio Janeiro.

197. Capsus semiochraceus.
Deræocoris semiochraceus, *Stal, Rio Jan. Hem*. 49.
Rio Janeiro.

198. Capsus caligatus.
Deræocoris caligatus, *Stal, Rio Jan. Hem*. 50.
Rio Janeiro.

199. Capsus testaceipes.
Deræocoris testaceipes, *Stal, Rio Jan. Hem*. 50.
Rio Janeiro.

200. Capsus semilotus.
Deræocoris semilotus, *Stal, Rio Jan. Hem*. 50.
Rio Janeiro.

201. CAPSUS CRIBRATUS.

Deræocoris cribratus, *Stal, Rio Jan. Hem.* 50.
Rio Janeiro.

202. CAPSUS LUCTUOSUS.

Deræocoris luctuosus, *Stal, Rio Jan. Hem.* 50.
Rio Janeiro.

203. CAPSUS PURGATUS.

Deræocoris purgatus, *Stal, Rio Jan. Hem.* 51.
Rio Janeiro.

204. CAPSUS STICTICOLLIS.

Deræocoris sticticollis, *Stal, Rio Jan. Hem.* 51.
Rio Janeiro.

205. CAPSUS STICTICUS.

Deræocoris sticticus, *Stal, Rio Jan. Hem.* 51.
Rio Janeiro.

206. CAPSUS CRIBROSUS.

Deræocoris cribrosus, *Stal, Rio Jan. Hem.* 51.
Rio Janeiro.

207. CAPSUS LENTICULOSUS.

Deræocoris lenticulosus, *Stal, Rio Jan. Hem.* 51.
Rio Janeiro.

208. CAPSUS DAHLBOMI.

Deræocoris Dahlbomi, *Stal, Rio Jan. Hem.* 52.
Rio Janeiro.

209. CAPSUS DILATATUS.

Deræocoris dilatatus, *Stal, Rio Jan. Hem.* 52.
Rio Janeiro.

210. CAPSUS FRAUDANS.

Deræocoris fraudans, *Stal, Rio Jan. Hem.* 52.
Rio Janeiro.

211. CAPSUS BASICORNIS.

Deræocoris basicornis, *Stal, Rio Jan. Hem.* 52.
Rio Janeiro.

212. Capsus cincticornis.

Deræocoris cincticornis, *Stal, Rio Jan. Hem.* 52.
Rio Janeiro.

213. Capsus vitreus.

Deræocoris vitreus, *Stal, Rio Jan. Hem.* 52.
Rio Janeiro.

214. Capsus clarus.

Deræocoris clarus, *Stal, Rio Jan. Hem.* 53.
Rio Janeiro.

215. Capsus insignis.

Deræocoris insignis, *Stal, Rio Jan. Hem.* 53.
Rio Janeiro.

Div. Calocoris.

216. Capsus bimaculatus.

Capsus bimaculatus, *Fabr. Syst. Rhyn.* 243—Calocoris bimaculatus, *Stal,
Hem. Fabr.* i. 86.
South America.

Div. Orthops.

217. Capsus Signoreti.

Signoreti, *Stal, Eug. Resa,* 257.
Rio Janeiro.

218. Capsus Bonariensis.

Bonariensis, *Stal, Eug. Resa,* 256.
Buenos Ayres.

Div. —?

219. Capsus speciosus.

speciosus, *Sgnt. A. S. E. F. 4me Sér.* iii. 571.
Chili.

220. Capsus ocellatus.

ocellatus, *Sgnt. A. S. E. F. 4me Sér.* iii. 572.
Chili.

Div. Resthenia.

A. Head black.
a. Head not pale beneath.
  * Head with a luteous or tawny hind spot. -   - circummaculatus.
 ** Head with no spot.

† Legs mostly black.

‡ Prothorax blackish. - - - - Zetterstedti.

‡‡ Prothorax pale. - - - - - pyrrhomelæna.

†† Legs mostly pale. - · - - - luteipes.

b. Head pale beneath. - - - - flavoniger.

B. Head blackish brown.

a. Abdomen striped.

* Prothorax blackish brown. - - - - subannulatus.

** Prothorax pale. - - - - - patruelis.

b. Abdomen not striped. - - - - bivittatus.

C. Head red or paler.

a. Corium wholly blackish.

* Prothorax not luteous in front.

† Second joint of the antennæ more than twice as long
as the first. - - - - - nigripennis.

†† Second joint of the antennæ hardly twice as long
as the first. - - - - seminiger.

** Prothorax luteous in front. - - - luteiceps.

b. Corium with a pale costa. - - - - costalis.

c. Corium olive-green. - - - - concinnus.

221. Capsus pyrrhula.

·Phytocoris pyrrhula, *Burm. Handb. Ent.* ii. 271.

a. Constancia. Presented by the Rev. H. Clark.

b, c. Rio Janeiro. Presented by J. Gray, Esq.

222. Capsus dimidiorufus.

Resthenia dimidiorufa, *Stal, Ofv. K. V. Ak. Forh.* xii. 186.
Brazil.

223. Capsus nigripennis.

Resthenia nigripennis, *Stal, Rio Jan. Hem.* 46.
Rio Janeiro.

224. Capsus Zetterstedti.

Resthenia Zetterstedti, *Stal, Rio Jan. Hem.* 46.
Rio Janeiro.

225. Capsus seminiger.

Resthenia seminigra, *Stal, Rio Jan. Hem.* 46.
Rio Janeiro.

226. Capsus pyrrhomelæna.

Resthenia pyrrhomelæna, *Stal, Rio Jan. Hem.* 46.
Rio Janeiro.

227. Capsus luteipes.

Resthenia luteipes, *Stal, Rio Jan. Hem.* 46.
Rio Janeiro.

228. Capsus flavoniger.

Resthenia flavonigra, *Stal, Rio Jan. Hem.* 46.
Rio Janeiro.

229. Capsus costalis.

Resthenia costalis, *Stal, Rio Jan. Hem.* 47.
Rio Janeiro.

230. Capsus concinnus.

Resthenia concinna, *Stal, Rio Jan. Hem.* 47.
Rio Janeiro.

231. Capsus subannulatus.

Resthenia subannulata, *Stal, Rio Jan. Hem.* 47.
Rio Janeiro.

232. Capsus bivittatus.

Resthenia bivittata, *Stal, Rio Jan. Hem.* 47.
Rio Janeiro.

233. Capsus patruelis.

Resthenia patruelis, *Stal, Rio Jan. Hem.* 47.
Rio Janeiro.

234. Capsus circummaculatus.

circummaculatus, *Stal, Ofv. K. V. Ak. Forh.* xi. 236; *Eug. Resa,* 257.
Buenos Ayres. Chili.

235. Capsus luteiceps.

luteiceps, *Stal, Eug. Resa,* 257.
Buenos Ayres.

236. Capsus Gayi.

Capsus melanochrus? *H.-Sch.*—Phytocoris Gayi, *Spin. Faun. Chil.* 184—
Lygæus picturatus, *Blanch. Faun. Chil.* 143—Capsus Gayi, *Sgnt. A. S. E. F.* 4me *Sér.* iii. 571.
Chili.

237. Capsus speciosus.

speciosus, *Sgnt. A. S. E. F.* 4*me Sér.* iii. 571..
Chili.

238. Capsus modestus.

modestus, *Blanch. Faun. Chil.* 187.   *Sgnt. A. S. E. F.* 4*me Sér.* iii. 572.
Chili.

239. Capsus elquiensis.

elquiensis, *Blanch. Faun. Chil.* 187.   *Sgnt. A. S. E. F.* 4*me Sér.* iii. 573.
Chili.

240. Capsus ocellatus.

ocellatus, *Sgnt. A. S. E. F.* 4*me Sér.* iii. 572.
Chili.

241. Capsus vicinus.

Phytocoris vicinus, *Blanch. Faun. Chil.* 186—Capsus vicinus, *Sgnt. A. S.*
    E. F. 4*me Sér.* iii. 573.
Chili.

242. Capsus tristis.

Phytocoris tristis, *Blanch. Faun. Chil.* 187—Capsus tristis, *Sgnt. A. S. E.*
    F. 4*me Sér.* iii. 573.
Chili.

243. Capsus antennatus.

Phytocoris antennatus, *Blanch. Faun. Chil.* 188—Capsus antennatus,
    Sgnt. A. S. E. F. 4*me Sér.* iii. 573.
Chili.

A. Corium deep black.
a. Head red.
  * Body smooth.   -   -   -   -   -   basalis.
** Body velvety.   -   -   -   -   -   velutinus.
b. Head black. -   -   -   -   -   tibialis.
B. Corium pale.
a. Corium streaked.   -   -   -   -   -   cinctipes.
b. Corium not streaked. -   -   -   -   -   xanthophilus.

244. Capsus basalis.

*Ater, ellipticus; caput et prothorax rufa, glabra; oculi prominuli;
    antennæ corporis dimidio non longiores; prothorax transverse bisul-
    catus; femora quatuor posteriora basi pallide flava.*

Deep black, elliptical. Head and prothorax bright red, smooth,
shining. Head triangular. Eyes black, prominent. Antennæ about half

the length of the body; first joint as long as the head; second much longer than the first; third shorter than the first; fourth shorter than the third. Prothorax with two triangular furrows near the fore border. Legs moderately long and slender; four hinder femora pale yellow towards the base. Length of the body 3—4 lines.

a—c. Constancia. Presented by the Rev. H. Clark.

245. CAPSUS TIBIALIS.

Ater, ellipticus, subtilissime punctatus, oculi prominuli ; rostrum luteum, coxas intermedias vix attingens ; antennæ corporis dimidio longiores ; prothorax luteus, antice ater ; pedes graciles, tibiis basi coxis femoribusque albis.

Deep black, elliptical, very finely punctured. Eyes piceous. Rostrum luteous, hardly extending to the middle coxæ. Antennæ more than half the length of the body; first joint as long as the head; second much longer than the first; third shorter than the first; fourth shorter than the third. Prothorax luteous, deep black in front of the transverse furrow, which is near the fore border. Legs slender; tibiæ towards the base, coxæ and femora white. Length of the body 3 lines.

a. Constancia. Presented by the Rev. H. Clark.
b. Petropolis. Presented by the Rev. H. Clark.

246. CAPSUS ATROLUTEUS.

Mas. Ater, fusiformis, velutinus, subtilissime punctatus; caput, prothorax, scutellum et pectus ochracea ; oculi subprominuli ; rostrum coxas posticas attingens; antennæ atræ, corpori æquilongæ ; prothoracis sulcus transversus anticus bene determinatus.

Female. Deep black, fusiform, velvety, very finely punctured. Head, prothorax, scutellum and pectus bright orange. Eyes piceous, slightly prominent. Rostrum piceous, extending to the hind coxæ. Antennæ deep black, slender, as long as the body; first joint a little longer than the head; second a little more than twice the length of the first; third more than half the length of the second; fourth shorter than the first. Prothorax with a distinct transverse furrow near the fore border. Legs slender. Length of the body 4 lines.

a. Rio Janeiro. Presented by W. W. Saunders, Esq.
b. Petropolis. Presented by J. Gray, Esq.

247. CAPSUS CINCTIPES.

Luteus, longi-ellipticus, glaber, nitens; oculi subprominuli ; rostrum coxas posticas attingens; antennæ nigræ, corpori æquilongæ, articulo 3o flavo apice flavescente, 4o fusco ; prothorax antice sulcatus, postice nigricante unifasciatus ; scutellum flavo univittatum ; pedes graciles, femoribus tibiisque fulvo bifasciatis ; corium flavo bistrigatum et unimaculatum ; membrana obscure cinerea, limpido bimaculata.

Luteous, elongate-elliptical, smooth, shining. Head triangular. Eyes black, slightly prominent. Rostrum extending to the hind coxæ. Antennæ black, slender, as long as the body; first joint as long as the head; second nearly twice the length of the first; third yellow, brownish at the tip, almost as long as the first; fourth brown, a little longer than the third. Prothorax with a variable blackish band on the hind border, and with a slight transverse furrow near the fore border. Scutellum with a yellow stripe. Legs slender; femora and tibiæ with two tawny bands; tarsi blackish at the tips. Corium with two yellow streaks and a subapical yellow spot; first streak extending from the base along the costa and thence obliquely to the disk; second streak on the interior border. Membrane dark cinereous, with two pellucid spots near the base. Length of the body 2½ lines.

a. Rio Janeiro. Presented by J. Gray, Esq.
b. Tejuca. Presented by the Rev. H. Clark.

248. Capsus xanthophilus.

Mas. Ferrugineus, glaber, nitens, fere linearis; caput nigrum; oculi prominuli; rostrum testaceum, coxas intermedias attingens; antennæ nigræ, pilosæ, corpore longiores, articulo 1o testaceo; prothorax antice arctatus et transverse sulcatus; scutellum testaceo univittatum; pedes rufi; corium apice luteum; membrana nigricans. Var.— Prothorax antice et corium picea; femora apice nigricantia.

Male. Ferruginous, smooth, shining, slender, nearly linear. Head black, short-triangular. Eyes prominent. Rostrum testaceous, extending to the middle coxæ; tip black. Antennæ black, pilose, longer than the body; first joint testaceous, stout, not longer than the head; second full six times the length of the first; third more than half the length of the second. Prothorax contracted and with a deep transverse furrow in front. Scutellum with a testaceous stripe. Legs red, slender. Corium luteous at the tip. Membrane and hind wings blackish. *Var. β.*—Prothorax in front and corium piceous. Femora blackish at the tips. Length of the body 3½—4½ lines.

a. Rio Janeiro. Presented by J. Gray, Esq.
b. Brazil. Presented by W. W. Saunders, Esq.

249. Capsus squalidus.

Mas. Luteus, ellipticus. subtilissime punctatus; capitis lobus intermedius antice niger; oculi prominuli; antennæ piceæ; prothorax transverse subsulcatus, postice piceo unifasciatus; corium apicem versus saturate rufum; alæ posticæ cinereæ.

Male. Luteous, elliptical, very finely punctured. Head triangular; middle lobe black in front. Eyes piceous, prominent. Antennæ piceous, slender; first joint as long as the head. Prothorax with a slight transverse furrow in front, and with a narrow piceous band on the hind border. Legs slender. Corium deep red towards the tip. Membrane brown. Hind wings cinereous. Length of the body 2 lines.

a. Rio Janeiro. Presented by W. W. Saunders, Esq.

250. Capsus incertus.

Luteus, fusiformis, subtilissime punctatus ; oculi subprominuli ; pro-thoracis margo subreflexus ; pedes longi ; membrana diaphana.

Luteous, fusiform, very finely punctured. Head triangular. Eyes piceous, slightly prominent. Prothorax with the fore border and the sides slightly reflexed. Legs long, slender. Membrane pellucid. Length of the body 3 lines.

a. Tejuca. Presented by the Rev. H. Clark.

251. Capsus obumbratus.

Luteus, ellipticus, nitens, subtilissime punctatus; facies nigro unistrigata ; oculi prominuli ; rostrum coxas intermedias attingens ; antennæ gracillimæ, corpore multo breviores, articulis 1o 2oque apices versus 3o 4oque nigricantibus ; prothorax nigro trimaculatus ; scutellum, femora postica et corium nigra ; membrana cinerea.

Luteous, elliptical, shining, very finely punctured. Head triangular ; a black streak on the face. Eyes black, prominent. Rostrum extending to the middle coxæ ; tip black. Antennæ very slender, much shorter than the body ; first and second joints blackish towards the tips ; second much longer than the others ; third and fourth blackish. Prothorax with a black spot on each side in front and with one near the hind border. Scutellum black. Legs slender ; hind femora black. Corium black. Membrane cinereous. Length of the body 1¾ line.

a. Petropolis. Presented by the Rev. H. Clark.

252. Capsus alternus.

Fœm.　*Rufus, longi-ellipticus, subtilissime punctatus ; oculi prominuli ; rostrum nigrum, coxas intermedias attingens ; antennæ nigræ, graciles ; prothorax litura nigra valde arcuata ; pedes graciles, femoribus apice tibiisque basi nigris ; corium fascia lata informi apiceque nigris ; membrana diaphana, nigro venosa.*

Female. Red, elongate-elliptical, very finely punctured. Head short-triangular. Eyes black, prominent. Rostrum black, extending to the middle coxæ. Antennæ black, slender ; second joint about thrice the length of the first. Prothorax with a black semicircular mark, the two ends joining the hind border. Legs slender ; femora black towards the tips ; tibiæ black towards the base. Corium with a broad irregular black band near the base and with a black tip. Membrane pellucid ; veins black. Length of the body 1½ line.

a. Rio Janeiro. Presented by W. W. Saunders, Esq.

253. Capsus leprosus.

Flavus, ovalis, tomentosus, fusco conspersus; rostrum coxas posticas attingens ; antennæ rufæ; prothorax nigro bipunctatus, marginem

*anticum versus transverse et tenuiter sulcatus; pedes rufi, flavo
conspersi, tibiis quatuor anterioribus albo unifasciatis, tarsis basi
albis; corium plagis paucis fuscis flavo guttatis; membrana
cinerea.*

Yellow, oval, tomentose. Head nearly triangular. Eyes black,
rather prominent. Rostrum extending to the hind coxæ. Antennæ red.
Prothorax and corium mottled with brown, this hue here and there forming
patches, which include yellow dots. Prothorax convex, with a black point
on each side of the disk in front and with a slight transverse furrow near
the fore border, its breadth much exceeding its length. Legs red, slender,
speckled with yellow; four anterior tibiæ with a white band beyond the
middle; tarsi white towards the base. Membrane cinereous. Length of
the body 2½ lines.

a. Santarem. From Mr. Bates' collection.

Galapagos.

254. CAPSUS SPOLIATUS.

*Pallide testaceus, fusiformis, subtilissime punctatus; oculi prominuli;
rostrum coxas posticas paullo superans; antennæ gracillimæ, corpori
æquilongæ; prothorax non sulcatus, lateribus non reflexis; corium
subhyalinum; membrana diaphana.*

Pale testaceous, fusiform, very finely punctured. Head triangular.
Eyes piceous, prominent. Rostrum extending a little beyond the hind
coxæ. Antennæ very slender, as long as the body; first joint as long as
the head; second more than twice the length of the first. Prothorax with
no transverse furrow; sides not reflexed. Legs slender. Corium nearly
hyaline. Membrane and hind wings pellucid. Length of the body
1 line.

a. Charles Isle, Galapagos. Presented by C. Darwin, Esq.
b. James Isle, Galapagos. Presented by C. Darwin, Esq.

255. CAPSUS NIGRITULUS.

*Niger, fusiformis, subtilissime punctatus; oculi prominuli; prothorax
antice subsulcatus; pedes testacei, graciles, femoribus nigricantibus;
corium piceum, costa strigaque basali testaceis, litura apicali nigra;
membrana fusca, cinereo marginata.*

Black, fusiform, very finely punctured. Head elongate-triangular.
Eyes piceous, prominent. Prothorax with a slight transverse furrow near
the fore border, which like the sides is slightly reflexed. Legs slender,
testaceous; femora blackish. Corium piceous; costa and a basal streak
testaceous; a black apical mark. Membrane brown, bordered with
cinereous. Length of the body 1¼ line.

a. Charles Isle. Presented by C. Darwin, Esq.

256. CAPSUS QUADRINOTATUS.

*Testaceus, fusiformis, nitens, subtilissime punctatus; capitis lobus
intermedius nigricans; oculi prominuli; rostrum coxas posticas
attingens; antennæ nigricantes, graciles, corpori æquilongæ, articulo
1o testaceo; prothorax postice piceo unifasciatus; scutellum piceo
bivittatum; pectus nigro biplagiatum; pedes gracillimi, femoribus
subincrassatis fusco trifasciatis basi albidis; corium fusco strigatum,
nigro biguttatum; membrana diaphana.*

Testaceous, fusiform, shining, very finely punctured. Head triangular;
middle lobe blackish. Eyes piceous, prominent. Rostrum extending to
the hind coxæ; tip black. Antennæ blackish, slender, as long as the body;
first joint testaceous, as long as the head; second more than twice as
long as the first; third much longer than the first; fourth shorter than
the third. Prothorax with a piceous band on the hind border. Scutellum
with two piceous stripes. Pectus with a black patch on each side. Legs
moderately long, very slender. Hind legs tawny; femora slightly
incrassated, with three irregular brown bands, whitish towards the base.
Corium slightly and diffusedly streaked with brown; two black dots on
the costa, one opposite the interior angle, the other apical. Membrane
pellucid. Length of the body 1½ line.

a—c. James Isle, Galapagos. Presented by C. Darwin, Esq.

Var. *Capitis lobus intermedius piceus; antennarum articulus 2us
testaceus, apice piceus; prothoracis fascia et scutelli vittæ pallide
fusca; macula costalis 1a fusca.*

Var. Middle lobe of the head piceous. Second joint of the antennæ
testaceous except towards the tip. Band of the prothorax and stripes of
the scutellum pale brown. Bands of the hind femora indistinct. First
costal spot brown.

d. Charles Isle, Galapagos. Presented by C. Darwin, Esq.

Africa.

A. Head with a longitudinal furrow. - - hottentottus
B. Head with no longitudinal furrow.
a. Eyes large.
  * Membrane speckled. - - - - straminicolor.
  ** Membrane not speckled. - - - obscuricornis.
b. Eyes of moderate size.
  * Third and fourth joints of the antennæ setaceous.
  † Membrane banded. - - - - ostentans.
  †† Membrane not banded.
  ‡ Body black. - - - - - histricus.
  ‡‡ Body piceous or pale.
  § Body not striped beneath.
  ✕ Scutellum not transversely furrowed.
  o Legs whitish. - - - - - capicola.
  oo Legs luteous, yellow or testaceous.

✦ Femora piceous. - - - - -		solitus.
✦✦ Femora not piceous.		
✚ Membrane brown. - - - -		sobrius.
✚✚ Membrane cinereous.		
∽ Head with a black spot. - - - -		illepidus.
∽∽ Head with no black spot.		
✛ Prothorax banded with brown. - - -		limbatus.
✛✛ Prothorax banded with red. - - -		suffusus.
✛✛✛ Prothorax not banded. - - - -		pallidulus.
✚✚✚ Membrane pellucid. - - - -		innotatus.
✕✕ Scutellum transversely furrowed. - -		conspersus.
§§ Body striped beneath. - - - -		sericeus.

257. Capsus capicola.

capicola, *Stal, Eug. Resa,* 256.

Cape.

Div.

Megacœlum, *Fieb. Stal, Hem. Afr.* iii. 18.

258. Capsus hottentottus.

Phythocoris hottentottus, *Stal, Ofv. Vet. Ak. Forh.* 1855, 36—Megacœlum hottentottum, *Stal, Hem. Afr.* iii. 18.

Caffraria.

259. Capsus ostentans.

Capsus ostentans, *Stal, Ofv. Vet. Ak. Forh.* 1855, 37—*Var.* Capsus histricus, *Stal, Ofv. Vet. Ak. Forh.* 1855, 37, Var. c and *d*—Deræocoris ostentans, *Stal, Hem. Afr.* iii. 20.

Cape. Caffraria.

260. Capsus histricus.

Capsus histricus, *Stal,| Ofv. Vet. Ak. Forh.* 1855, 37, Var. *a* and *b.*—Deræocoris histricus, *Stal, Hem. Afr.* iii. 21,

Caffraria.

261. Capsus incomparabilis.

Capsus incomparabilis, *Stal, Ofv. Vet. Ak. Forh.* 1855, 35—Deræocoris incomparabilis, *Stal, Hem. Afr.* iii. 212.

Caffraria.

Div.

Volumnus, *Stal, Hem. Afr.* iii. 19.

262. CAPSUS STRAMINICOLOR.

Capsus straminicolor, *Stal, Ofv. Vet. Ak. Forh.* 1855, 36—Volumnus
straminicolor, *Stal, Hem. Afr.* iii. 19.
Caffraria.

263. CAPSUS OBSCURICORNIS.

Capsus obscuricornis, *Stal, Ofv. Vet. Ak. Forh.* 1855, 36—Volumnus
obscuricornis, *Stal, Hem. Afr.* iii. 19.
Caffraria.

264. CAPSUS SOBRIUS.

Fœm. *Ferrugineus, fusiformis, subtilissime punctatus, subtus luteus;
caput luteum; oculi subprominuli; rostrum coxas intermedias
attingens; antennæ nigræ, graciles,* corpore *paullo breviores; prothorax antice nigro bipunctatus. postice piceo bimaculatus, lateribus
nigris; pedes lutei, validi; corium costa guttaque discali nigris,
striga apicali alba; membrana fusca.*

Female. Ferruginous, fusiform, very finely punctured, luteous beneath.
Head luteous, triangular. Eyes black, slightly prominent. Rostrum
extending to the middle coxæ. Antennæ black, slender, a little shorter
than the body; first joint much longer than the head; second a little more
than twice as long as the first; third longer than the first. Prothorax
with two large piceous spots on the hind border and with two black points
in the fore part; fore border and sides reflexed, the latter black. Legs
luteous, rather stout. Corium with a black costa, with a black dot in the
disk near the exterior border and with a white apical streak. Membrane
brown. Length of the body 3 lines.

a. Sierra Leone. Presented by the Rev. D. F. Morgan.

265. CAPSUS ILLEPIDUS.

Mas. *Fulvus, fusiformis, subtilissime punctatus, subtus flavus; facies
nigro unimaculata; oculi subprominuli; rostrum coxas intermedias
attingens; antennæ fuscæ, graciles,* corpore *paullo breviores; prothorax antice bicallosus; pedes longi, graciles; membrana pallide
cinerea.*

Male. Tawny, fusiform, very finely punctured, yellow beneath. Head
triangular; a black spot on the face. Eyes slightly prominent. Rostrum
extending to the middle coxæ; tip black. Antennæ brown, slender, a
little shorter than the body; first joint longer than the head; second more
than twice as long as the first; third as long as the first; fourth much
shorter than the third. Prothorax with a callus on each side in front.
Legs long, slender. Membrane pale cinereous. Hind wings pellucid.
Length of the body 3 lines.

a. Cape. From M. Dregé's collection.

266. Capsus solitus.

Piceus, fusiformis, subtilissime punctatus, cinereo tomentosus; oculi prominuli; antennæ flavæ, graciles, corpore paullo breviores, articulis 2o 3oque apice 1oque nigris; prothorax antice subsulcatus; pedes flavi, femoribus piceis basi flavis; membrana obscure cinerea.

Piceous, fusiform, very finely punctured, with cinereous tomentum. Head elongate-triangular. Eyes prominent. Antennæ yellow, slender, a little shorter than the body; first joint black, a little shorter than the head; second black towards the tip, more than twice as long as the first; third black at the tip, as long as the first. Prothorax with a slight transverse furrow near the fore border. Legs yellow, slender; femora piceous, yellow at the base. Membrane dark cinereous. Length of the body 2 lines.

a. Cape. From M. Dregé's collection.

267. Capsus pallidulus.

Pallide flavus, fusiformis, subtilissime punctatus; oculi subprominuli; prothoracis sulco transverso antico indeterminato; membrana cinerea.

Pale yellow, fusiform, very finely punctured. Head elongate-triangular. Eyes piceous, slightly prominent. Antennæ slender; first joint as long as the head; second more than twice as long as the first. Prothorax with a slight transverse furrow in front. Legs slender. Membrane cinereous. Hind wings pellucid. Length of the body 1½ line.

a. Cape. From M. Dregé's collection.

268. Capsus conspersus.

Fœm. Flavus, ellipticus, subtilissime punctatus, subtus nitens; eaput valde convexum; oculi subprominuli; prothorax luteo bistrigatus, antice bicallosus; scutellum transverse sulcatum, luteo biguttatum et bistrigatum; corii venæ luteo marginatæ; membrana pallide cinerea.

Female. Yellow, elliptical, very finely punctured, shining beneath. Head triangular; vertex very convex. Eyes piceous, slightly prominent, Prothorax with two exterior longitudinal streaks in the disk and with a strongly-marked callus on each side in front; sides reflexed. Scutellum wi;h a transverse furrow, with two anterior luteous dots and with two posterior luteous streaks. Legs stout. Corium luteous along the veins. Membrane pale cinereous. Hind wings pellucid. Length of the body 2 lines.

a. Cape. From M. Dregé's collection.

269. Capsus innotatus.

Fœm. Testaceus, ellipticus, nitens, subtilissime punctatus; oculi subprominuli; rostrum coxas posticas attingens; antennæ graciles; prothorax antice bicallosus; membrana diaphana.

Female. Testaceous, elliptical, shining, very finely punctured. Head elongate-triangular. Eyes piceous, slightly prominent. Rostrum extending to the hind coxæ; tip black. Antennæ slender; first joint as long as the head; second much more than twice as long as the first; third longer than the first. Prothorax with a callus on each side in front. Legs slender. Membrane pellucid. Length of the body 2½ lines.

a. Interior of South Africa. Presented by the Earl of Derby in 1843.

270. Capsus suffusus.

Fœm. *Testaceus, fusiformis, subtilissime punctatus; oculi subprominuli; rostrum coxas posticas attingens; antennæ graciles,* corpore breviores; *prothorax rufo postice unifasciatus; scutellum apice rufum; corium rufum, vitta costali testacea; membrana cinerea.*

Female. Testaceous, fusiform, very finely punctured. Head triangular. Eyes piceous, slightly prominent. Rostrum extending to the hind coxæ; tip black. Antennæ slender, shorter than the body; first joint as long as the head; second more than twice as long as the first. Prothorax with a red band on the hind border. Scutellum red towards the tip. Legs slender. Corium red, with a testaceous costal streak extending from the base to half the length. Membrane cinereous. Length of the body 3½ lines.

a. Cape. From M. Dregé's collection.

271. Capsus limbatus.

Testaceus, fusiformis, subtilissime punctatus; oculi subprominuli; antennæ graciles, corpore breviores, *articulo 2o apice nigro; prothorax antice bicallosus, postice fusco unifasciatus; corium fusco univittatum et unimaculatum; membrana cinerea.*

Testaceous, fusiform, very finely punctured. Head short-triangular. Eyes piceous, slightly prominent. Antennæ slender, shorter than the body; first joint as long as the head; second black at the tip, a little more than twice as long as the first; third longer than the first. Prothorax with a callus on each side in front and with a brown band on the hind border. Legs slender. Corium with a brown stripe along the interior border and with a brown spot on the interior angle. Membrane cinereous. Length of the body 1¾ line.

a. Cape. From M. Dregé's collection.

272. Capsus sericeus.

Fœm. *Piceus, fusiformis, subtilissime punctatus, cinereo tomentosus, subtus testaceus piceo bivittatus; caput fulvum, antice piceum; oculi prominuli; rostrum testaceum, coxas posticas paullo superans; antennæ testaceæ,* corpore breviores; *prothorax antice fulvus et bicallosus; pedes picei; corium apice testaceum; membrana obscure cinerea.*

Female. Piceous, fusiform, very finely punctured, with cinereous tomentum, testaceous beneath. Head tawny, elongate-triangular, piceous in front. Eyes piceous, prominent. Rostrum testaceous, extending a little beyond the hind coxæ. Antennæ testaceous, slender, shorter than the body; first joint a little longer than the head; second more than twice as long as the first; third longer than the first. Prothorax tawny towards the fore border, near which there is a callus on each side. Pectus and under side of the abdomen with a piceous stripe on each side. Legs piceous, slender. Corium testaceous at the tip. Membrane dark cinereous. Length of the body 2 lines.

a. Sierra Leone. Presented by the Rev. D. F. Morgan.

South Asia.

A. Head with no whitish spots.
  a. Eyes not large.
    * Rostrum extending to the middle coxæ.   -   fasciatus.
    ** Rostrum extending beyond the middle coxæ. - vitripennis.
    *** Rostrum extending to the hind coxæ.
    † Head streaked. - - - - partitus.
    †† Head not streaked.
    ‡ Body pale.
    § Prothorax not banded. - - - - stramineus.
    §§ Prothorax banded. - - - - Sinicus.
    §§§ Prothorax with a piceous disk. - - - discoidalis.
    ‡‡ Body black. - - - - - patulus.
    **** Rostrum extending somewhat beyond the hind coxæ. - - - - lineifer
    ***** Rostrum extending to the tip of the abdomen. - canescens.
    ****** Rostrum extending beyond the tip of the abdomen. vicarius.
  b. Eyes large. - - - - - incisus.
B. Head with two whitish spots. - - - Chinensis.

273. Capsus Chinensis.

Chinensis, *Stal, Eug. Resa,* 258.

Hong Kong.

Div.

Hyalopeplus, *Stal, Ofv. K. V. Ak. Forh.* 1870, 670.

274. Capsus vitripennis.

Capsus vitripennis, *Stal, Eug. Resa, Hem.* 255—Hyalopeplus vitripennis, *Stal, Ofv. K. V. Ak. Forh.* 1870, 671.

Malacca. Java. Philippine Isles.

275. Capsus semiclusus.

Niger, fusiformis, subtilissime punctatus; caput longiusculum; oculi prominuli; antennarum articulus 1us *capiti æquilongus,* 2us *subclavatus* 1o *plus duplo longior; scutellum sat magnum, apice ferrugineum; membrana cinerea, nigro venosa, nigricante submarginata.*

Black, fusiform, dull, very finely punctured. Head elongate-triangular. Eyes prominent. First joint of the antennæ as long as the head; second subclavate, more than twice as long as the first. Prothorax with a slight callus on each side behind the well-defined transverse furrow, which is near the fore border. Scutellum rather large, ferruginous at the tip. Membrane cinereous, blackish-bordered along part of the exterior border; veins black; second areolet more than four times the size of the first; a whitish mark between it and a short black line. Length of the body 1¾ line.

a. Ceylon. Presented by Dr. Thwaites.

276. Capsus partitus.

Mas. *Flavus, fusiformis, nitens, subtilissime punctatus; caput luteum, nigro unistrigatum; oculi subprominuli; rostrum coxas posticas attingens; antennæ nigræ, gracillimæ, corpore paullo breviores, articulis 2o basi 1oque flavis; prothorax antice lutescens; pectus nigro bimaculatum; ventris discus niger; corium fascia obliqua margine interiore apiceque nigris; membrana cinerea.*

Male. Yellow, fusiform, shining, very finely punctured. Head luteous, short-triangular; a black streak on the front. Eyes black, slightly prominent. Rostrum extending to the hind coxæ. Antennæ black, very slender, a little shorter than the body; first joint yellow, a little longer than the head; second much more than twice as long as the first, yellow towards the base; third much longer than the first. Prothorax with a luteous tinge in front. Pectus with a black spot on each side. Abdomen beneath with a black disk. Corium black along the interior border, with a black oblique band beyond the middle; tip black. Membrane cinereous. Length of the body 2 lines.

a. North Hindostan. From Capt. Boyes' collection.

277. Capsus subirroratus.

Niger, ellipticus, subtiliter punctatus; oculi subprominuli; antennæ corpore breviores; prothorax valde convexus, lateribus vix reflexis; scutellum apice fulvum; corium fulvo punctatum et triguttatum; membrana nigricans.

Black, elliptical, dull, finely punctured. Head small. Eyes slightly prominent. Antennæ slender, shorter than the body; first joint moderately stout, a little longer than the head. Prothorax very convex; sides hardly reflexed. Tip of the scutellum tawny. Corium with a few tawny points and with three tawny dots, one at the base, one in the exterior part of the disk, and one at the tip. Membrane blackish. Length of the body 1½ line.

a. Ceylon. Presented by Dr. Thwaites.

278. Capsus stramineus.

Mas. *Pallide flavus, fusiformis, subtilissime punctatus; oculi promi-*
nuli; rostrum coxas posticas attingens; antennæ graciles, corpore
longiores; prothorax antice bicallosus, lateribus perparum reflexis;
pedes longi, gracillimi; membrana pallide cinerea.

Male. Pale yellow, fusiform, very finely punctured. Head triangular.
Eyes black, prominent. Rostrum extending to the hind coxæ; tip black.
Antennæ slender, longer than the body; first joint longer than the head;
second much more than twice as long as the first; third a little more
than twice as long as the first; fourth as long as the first. Prothorax
with a callus on each side in front; sides very slightly reflexed. Legs long,
very slender. Membrane pale cinereous. Length of the body 2½ lines.

a. North Bengal. From Lieut. Campbell's collection.

279. Capsus patulus.

Fœm. *Niger, brevis, nitens, subtilissime punctatus; caput rufum;*
rostrum coxas posticas attingens; antennæ luteæ, corpore breviores.

Female. . Black, short-elliptical, shining, very finely punctured. Head
red, triangular. Rostrum red, extending to the hind coxæ. Antennæ
luteous, shorter than the body; first joint shorter than the head; second
black towards the tip, more than twice as long as the first. Prothorax
with a lobe on each side in front and with no transverse furrow; hind
border ferruginous. Disk of the pectus red. Legs red, slende; femora
with a black hand beyond the middle; hind femora black at the base.
Membrane brownish cinereous. · Length of the body 2 lines.

a. North Hindostan. From Capt. Boyes' collection.

280. Capsus Sinicus.

Mas. *Testaceus, fusiformis, subtilissime punctatus; oculi prominuli;*
rostrum coxas posticas attingens; antennæ corpore æquilongæ; pro-
thorax antice bilobatus, postice fusco unifasciatus; pedes longi,
graciles, femoribus posticis apice nigris; membrana cinerea, nigricante
venosa. Fœm.—Rufescens; scutellum et corium fusca, hujus costa
rufescens.

Male. Testaceous, fusiform, very finely punctured. Head triangular.
Eyes piceous, prominent. Rostrum extending to the hind coxæ. Antennæ
slender, as long as the body; first joint longer than the head; second more
than twice as long as the first. Prothorax with a lobe on each side in
front, and with a brown band on the hind border. Legs long, slender;
hind femora with black tips. Membrane cinereous; veins blackish.
Female.—Reddish. Scutellum and corium brown, the latter with a reddish
costa. Length of the body 2½ lines.

a, b. Hong Kong. Presented by J. C. Bowring, Esq.

281. Capsus vicarius.

Mas. *Fuscus, fusiformis, subtilissime punctatus, subtus testaceus; caput longiusculum; oculi prominuli; rostrum abdominis apicem paullo superans; antennæ gracillimæ; prothorax antice arctatus et transverse subsulcatus, postice testaceo tristrigatus; pectus nigro bimaculatum; venter fasciis duabus interruptis apiceque nigris; pedes longi, gracillimi, femoribus testaceo trifasciatis, tibiis testaceo unifasciatis; corium pustulis strigis_ duabus macula apiceque albidis; membrana cinerea.*

Male. Brown, fusiform, very finely punctured, testaceous beneath. Head elongate-triangular. Eyes black, prominent. Rostrum extending a little beyond the tip of the abdomen. Antennæ very slender; first joint as long as the head; second more than twice as long as the first. Prothorax contracted in front and with a slight transverse furrow; three testaceous streaks on the hind border. Pectus with a black spot on each side. Abdomen beneath with two interrupted black bands and with a black tip; first band very broad. Legs long, very slender; femora with three testaceous bands; tibiæ with a testaceous band. Corium with minute whitish pustules, with two whitish streaks on the interior border, with a whitish spot in the disk and with a whitish tip. Membrane cinereous. Length of the body 3½ lines.

a. Siam. Presented by W. W. Saunders, Esq.

282. Capsus incisuratus.

Fuscus, fusiformis, subtus testaceus; caput breve; oculi magni, prominuli; antennæ testaceæ, gracillimæ, corpori æquilongæ; prothorax non sulcatus; pedes testacei; corium apud costam flavescente bistrigatum.

Brown, fusiform, testaceous beneath. Head short-triangular. Eyes black, large, prominent. Antennæ testaceous, very slender, as long as the body; first joint longer than the head; second very much longer than the first; third longer than the first. Prothorax with no transverse furrow. Legs testaceous, slender. Corium with two pale yellowish hyaline costal streaks. Membrane cinereous. Length of the body 1½ line.

a. Ceylon. Presented by Dr. Thwaites.

283. Capsus canescens.

Fœm. *Fuscus, fusiformis, subtiliter punctatus, cano tomentosus; caput longiusculum; oculi magni; rostrum abdominis apicem attingens; antennæ gracillimæ, corpore longiores; prothorax transverse subsulcatus; corium albido tuberculatum, fasciola apiceque albis; membrana fusca, albo bistrigata.*

Female. Brown, fusiform, finely punctured, with hoary tomentum. Head elongate-triangular. Eyes large. Rostrum extending to the tip of the abdomen. Antennæ very slender, longer than the body; first joint a

little shorter than the head; second more than twice as long as the
first; third and fourth as long as the second. Prothorax with a slight
transverse furrow before the middle. Corium with minute whitish tubercles,
with an incomplete white band beyond the middle and with a white tip.
Membrane brown, with two white streaks in each. Length of the body
3½ lines.

a. Malacca.   Presented by W. W. Saunders, Esq.

284. Capsus fasciatus.

Fœm.   *Rufus, nitens, fere linearis; caput parvum; oculi valde prominuli; rostrum coxas intermedias attingens; antennæ nigræ, graciles,
hirsutulæ; prothorax antice arctatus et transverse bisulcatus; pedes
graciles, hirsutuli, femoribus subtus tibiis apice tarsisque pallide
flavis; alæ fuscæ, nigricante venosæ.*

Female. Red, smooth, shining, slender, nearly linear. Head small,
short. Eyes black, very prominent. Rostrum extending to the middle
coxæ. Antennæ black, slender, minutely hirsute; first joint rather stout,
as long as the head; second more than thrice as long as the first.
Prothorax contracted in front, where there are two well-defined transverse
furrows. Legs slender, minutely hirsute; femora beneath, tibiæ towards
the tips and tarsi pale yellow. Wings brown, slightly hyaline; veins
blackish. Length of the body 4 lines.

a. Singapore.   Presented by W. W. Saunders, Esq.

285. Capsus lineifer.

Mas.   *Testaceus, subtilissime punctatus, nigro aut rufo quinque vittatus;
oculi prominuli; rostrum coxas posticas sat superans; antennæ piceæ,
corpori æquilongæ; prothorax vix sulcatus; pedes longiusculi, sat
graciles; alæ cinereæ, piceo venosæ.*

Male. Testaceous, fusiform, very finely punctured. Head and prothorax with three black or red parallel lines, the lateral pair abbreviated
on the prothorax hindward. Head triangular. Eyes piceous, prominent.
Rostrum extending somewhat beyond the hind coxæ. Antennæ piceous,
slender, as long as the body; first joint stout, as long as the head; second
more than thrice as long as the first and less than thrice as long as
the third; fourth shorter than the third. Prothorax with two exterior
black or red lines on each side, in addition to those before mentioned;
transverse furrow extremely slight. Legs rather long and slender. Wings
cinereous; veins piceous. Length of the body 4 lines.

a, b. Malacca.   Presented by W. W. Saunders, Esq.

286. Capsus discoidalis.

Fœm.   *Testaceus, subtilissime punctatus; oculi prominuli; rostrum
coxas posticas attingens; antennæ piceæ, corpore longiores, basi
testaceæ, articulo 3o basi albo; prothorax transverse subsulcatus,
disco piceo; pedes longiusculi; corium fusco marginatum; membrana
cinerea, piceo venosa.*

Female. Testaceous, fusiform, very finely punctured. Head short. Eyes piceous, prominent. Rostrum extending to the hind coxæ; tip black. Antennæ piceous, slender, somewhat longer than the body; first joint a little longer than the prothorax, testaceous at the base; second very much longer than the first; third a little shorter than the second, white towards the base; fourth less than half as long as the third. Prothorax piceous except in front and along each side; a slight transverse furrow near the fore border. Legs slender, rather long. Corium brown along the interior border and along the exterior border. Membrane cinereous; veins piceous. Length of the body 4 lines.

a. Malacca. Presented by W. W. Saunders, Esq.
b. Singapore. Presented by W. W. Saunders, Esq.

Australasia.

A. Body black or piceous.
a. Membrane black. - - - - - tristis.
b. Membrane brown; veins with cinereous borders. - simulans.
c. Membrane cinereous, brown at the tip.
 * Scutellum yellow at the tip. - - - - apicifer.
 ** Scutellum not yellow at the tip. - - - costalis.
d. Membrane cinereous.
 * Prothorax black. - - - - - pellucidus.
 ** Prothorax yellowish. - - - - - collaris.
B. Body red. - - - - - - lucidus.
C. Body whitish yellow. - - - - - Tagalicus.

287. Capsus Tagalicus.

Tagalicus, *Stal, Eug. Resa,* 258.
Manilla.

Div.

Helopeltis, *Sgnt.*

288. Capsus pellucidus (bis lectum).

Helopeltis pellucida, *Stal, Ofv. K. V. Ak. Forh.* 1870, 667.
Philippine Isles.

289. Capsus collaris (bis lectum).

Helopeltis collaris, *Stal, Ofv. K. V. Ak. Forh.* 1870, 667.
Philippine Isles.

Div.

Eurystylus, *Stal, Ofv. K. V. Ak. Forh.* 1870, 671.

290. Capsus costalis.

Eurystylus costalis, *Stal, Ofv. K. V. Ak. Forh.* 1870, 671.
Philippine Isles.

Div.

Macralonius, *Stal, Ofv. K. V. Ak. Forh.* 1870, 670.

291. Capsus sobrinus.

Macralonius sobrinus, *Stal, Ofv. K. V. Ak. Forh.* 1870, 670.
Java.

292. Capsus thoracatus.

thoracatus, *Stal, Ofv. K. V. Ak. Forh.* 1870, 671.
Java.

293. Capsus apicifer.

Fœm.   *Niger, opacus, subtilissime punctatus; oculi subprominuli; rostrum coxas intermedias attingens; antennæ corpori æquilongæ, articulo 2o subclavato; prothorax transverse subsulcatus; scutellum apice flavum; pectus testaceo bifasciatum; membrana cinereo-hyalina, apice fusca, nigricante venosa.*

Female. Black, elliptical, dull, very finely punctured. Head triangular, small. Eyes slightly prominent. Rostrum extending to the middle coxæ. Antennæ as long as the body; first joint elongate-fusiform, longer than the head; second subclavate, very much longer than the first. Prothorax with a very slight transverse furrow in front. Scutellum yellow at the tip. Pectus with two dull testaceous bands. Legs rather stout. Membrane hyaline, cinereous, brown at the tip; veins blackish. Hind wings cinereous-hyaline. Length of the body 2½ lines.

a. Makian, Celebes.   Presented by W. W. Saunders, Esq.

294. Capsus lucidus.

Mas.   *Rufus, subtiliter punctatus, fere linearis, subtus flavus; vertex piceus; oculi prominuli; rostrum coxas posticas paullo superans; antennæ nigræ, gracillimæ, corpore longiores, articulo 3o basi albo; thorax piceus, rufescente marginatus; abdomen apice nigrum; pedes rufi; corium flavescente vitreum, coccineo sexstrigatum; membrana pallide fusca, fascia obscuriore.*

Male. Red, slender, shining, finely punctured, nearly linear, yellow beneath. Head short, triangular; vertex piceous. Eyes piceous, prominent. Rostrum extending a little beyond the hind coxæ. Antennæ black, very slender, longer than the body; first joint nearly as long as the prothorax; second much more than twice as long as the first; third white towards the base, longer than the first; fourth shorter than the first. Prothorax and scutellum piceous, the former reddish on each side, the latter reddish at the tip. Abdomen black at the tip. Legs and corium nearly hyaline, with a yellowish tinge, and on each side with three crimson streaks, which radiate from the end of the interior side. Membrane pale brown, with a darker band. Length of the body 2¼ lines.

a. Sarawak. Presented by W. W. Saunders, Esq.

295. Capsus simulans.

Fœm. *Piceus, crassus, subtiliter punctatus, subtus testaceus; caput rufum; oculi prominuli; rostrum coxas posticas attingens; antennæ gracillimæ, corpore paullo breviores; prothorax antice rufus et transverse sulcatus, postice rufescente univittatus; pedes picei; corium cinereo venosum; membrana fusca, ven. cinereo marginatis.*

Female. Piceous, elliptical, thick, finely punctured, testaceous beneath. Head and fore part of the prothorax red. Head transverse. Eyes black, prominent. Rostrum extending to the hind coxæ. Antennæ very slender, a little shorter than the body; first joint as long as the head; second very much longer than the first; third a little longer than the second; fourth longer than the third. Prothorax with a well-defined transverse furrow before the middle. Scutellum and bind part of the prothorax with a reddish stripe, the latter reddish at the tip. Legs piceous. Corium with cinereous veins. Membrane brown; veins bordered with cinereous. Length of the body 3 lines.

a. Singapore. Presented by W. W. Saunders, Esq.

296. Capsus tristis.

Niger, gracilis, fere linearis, subtiliter punctatus; caput testaceum; oculi prominuli; antennæ corpore breviores; prothorax antice arctatus, sulco transverso medio bene determinato; pectoris discus albus, rufo biguttatus; femora basi, tibiæ apice tarsique alba. Var.—*Prothorax ferrugineus; femora testacea, piceo bifasciata.*

Black, slender, nearly linear, finely punctured. Head testaceous, triangular. Eyes piceous, prominent. Antennæ rather stout, shorter than the body; first joint as long as the head; second more than twice as long as the first. Prothorax elongate, contracted in front, with a well-defined transverse middle furrow. Disk of the pectus white, with a red dot on each side. Legs slender; tarsi, femora towards the base and tibiæ towards the tips white. *Var.*—Immature? Prothorax ferruginous. Femora testaceous, with two piceous bands. Length of the body 3 lines.

a, b. New Guinea. Presented by W. W. Saunders, Esq.

Australia.

A. Membrane with red-bordered veins. -   -   -   Kinbergi.
B. Membrane with no red-bordered veins.
  a. Body whitish yellow or pale green.
    * Corium without red-bordered veins.
      † First joint of the antennæ longer than the head. --   dilutus.
     ††† First joint of the antennæ shorter than the head. --   Sidnicus.
     ††† First joint of the antennæ as long as the head.

‡ Body striped.   -   -   -   -   -   angulifer.
‡‡ Body not striped. -   -   -   -   -   intaminatus.
** Corium with red-bordered veins.   -   -   -   pictulifer.
b. Body cinnamon colour.   -   -   -   -   Dallasi.

297. Capsus varicornis.

Phytocoris varicornis, *Erichs. Wieg. Arch.* viii. 280.
Tasmania.

298. Capsus Dallasi.

Dallasi, *Stal, Eug. Resa,* 258.
Sydney.

299. Capsus Kinbergi.

Kinbergi, *Stal, Eug. Resa,* 255.
Sydney.

300. Capsus dilutus.

dilutus, *Stal, Eug. Resa,* 256.
Sydney.

301. Capsus Sidnicus.

Sidnicus, *Stal, Eug. Resa,* 258.
Sydney.

302. Capsus angulifer.

Fœm. *Testaceus, gracilis, fere linearis, subtilissime punctatus; oculi
subprominuli; rostrum coxas posticas attingens; antennæ corpore
æquilongæ, articulis 2o 3oque apices versus nigris; prothorax piceo
bivittatus; membrana fusca.*

Female. Testaceous, narrow, nearly linear, very finely punctured.
Head triangular. Eyes reddish, slightly prominent. Rostrum extending
to the hind coxæ; tip black. Antennæ as long as the body; first joint as
long as the head; second nearly thrice as long as the first; black
towards the tip; third nearly twice as long as the first, black for half
the length from the tip. Prothorax with two oblique piceous stripes,
which converge to the fore border, where they are united. Legs slender.
Membrane brown. Length of the body 2½ lines.

a. Australia. Presented by the Haslar Hospital.

303. Capsus pictulifer.

Mas. *Pallide testaceus, fusiformis, subtilissime punctatus, rufo quadrivittatus; oculi prominuli; antennæ corpore paullo breviores; scutelli
discus rufus; pectus rufo biplagiatum; alæ cinereæ, corii venis rufo
marginatis.*

Male. Pale testaceous, fusiform, very finely punctured, paler beneath.
Head and prothorax with two red parallel stripes, which are united in front

of the eyes. Head short-triangular. Eyes piceous, prominent. Antennæ slender, a little shorter than the body ; first joint rather stout, a little longer than the head; second nearly thrice as long as the second, and more than twice as long as the third. Prothorax with a transverse impression before the middle, and on each side with a red stripe, which is parallel to the border. Scutellum with a red disk. Pectus with a red patch on each side. Legs slender. Wings cinereous. Corium with red-bordered veins. Length of the body 2¼ lines.

a. Australia. Presented by the Entomological Club.

304. CAPSUS INTAMINATUS.

Mas. Testaceus, fusiformis, subtilissime punctatus; facies nigricans; oculi subprominuli ; rostrum coxas posticas attingens ; antennæ corpore paullo breviores; prothorax antice bicallosus, membrana albida.

Male. Testaceous, fusiform, very finely punctured. Head short-triangular; face blackish. Eyes piceous, slightly prominent. Rostrum extending to the hind coxæ ; tip black. Antennæ slender, a little shorter than the body; first joint rather stout, as long as the head ; second more than twice as long as the first; third much shorter than the second. Prothorax with a callus, and a slight transverse hinder furrow on each side in front. Legs slender; femora rather stout. Membrane whitish. Length of the body 2½ lines.

a, b. Australia.

Var. ?—Membrane cinereous. Length of the body 2 lines.

c, d. Tasmania. Presented by W. W. Saunders. Esq.

Oceania.
305. CAPSUS TAITICUS.
Taiticus, *Stal, Eug. Resa*, 257.
Tahiti.

306. CAPSUS PACIFICUS.
pacificus, *Stal, Eug. Resa*, 256.
Tahiti.

307. CAPSUS PELLUCIDUS (bis lectum).
pellucidus, *Stal, Eug. Resa*, 255.
Honolulu.

308. CAPSUS LATICINCTUS.

Fœm. Fulvus, subtilissime punctatus, subtus nigro biplagiatus et bivittatus ; oculi prominuli; rostrum coxas posticas attingens; antennæ corpore paullo breviores, apices versus piceæ ; prothorax nigro interrupte quadrivittatus; corium apud marginem interiorem piceum, venis ex parte rufis ; membrana fuscescens.

Female. Tawny, elliptical, very finely punctured. Head triangular. Eyes piceous, prominent. Rostrum extending to the hind coxæ; tip black. Antennæ slender, a little shorter than the body; first joint a little shorter than the head; second about twice as long as the first, piceous towards the tip; third piceous, much shorter than the second. Prothorax with two interrupted black stripes on each side, the inner pair continued on each side of the scutellum. Pectus and basal part of the abdomen with a black patch in the middle and with a black stripe on each side. Corium irregularly piceous along the interior border; veins partly red. Membrane brownish. Length of the body 2½ lines.

a. New Zealand. Presented by Dr. Sinclair.

309. Capsus ustulatus.

Fœm. *Fulvus, subtilissime punctatus; oculi prominuli; rostrum coxas intermedias attingens; antennæ corpore paullo breviores, apices versus piceæ; prothorax vittis quinque pallidioribus duabusque piceis; scutellum pallido univittatum; pectoris discus et venter basi nigricantes; femora postica subtus picea; membrana fuscescens, cinereo guttata.*

Female. Tawny, elliptical, very finely punctured. Head triangular, with some smaller paler marks. Eyes piceous, prominent. Rostrum extending to the middle coxæ; tip black. Antennæ slender, a little shorter than the body; first joint stout, a little shorter than the head; second about half as long as the first, piceous at the tip; third and fourth piceous; third a little more than half as long as the second; fourth shorter than the third. Prothorax with five paler stripes, the middle one more slender than the other four, which are abbreviated and include on each side an abbreviated piceous stripe. Scutellum with a paler stripe. Disk of the pectus and abdomen beneath towards the base blackish. Legs slender; hind femora piceous beneath. Membrane brownish, with several pale cinereous dots. Length of the body 2½ lines.

a. New Zealand. Presented by Dr. Sinclair.

Country unknown.

310. Capsus marginicollis.

Fœm. *Niger, fusiformis, opacus; caput parvum; antennæ graciles; prothorax antice et utrinque ochraceus, sulco transverso bene determinato; propectoris latera ochracea; membrana nigricans.*

Female. Black, fusiform, dull. Head small, short-triangular. Eyes slightly prominent. Antennæ slender; first joint a little longer than the head; second a little more than twice as long as the first. Prothorax ochraceous on each side and towards the fore border, near which there is a well-defined transverse furrow. Prothorax ochraceous along each side. Legs slender. Membrane blackish. Length of the body 3 lines.

a. ——?

Genus 6. LEPTOMEROCORIS.

Leptomerocoris, *Kirschb. Caps.* 31.

Europe, West Asia and Siberia.

Div. 1.

Halticus, *Hahn. Wanz. Ins.* i. 113. *Fieb. Crit. Gen.* 47; *Eur. Hem.* 69, 281—Halticocoris, *Dougl. and Scott, Hem.* 478.

1. LEPTOMEROCORIS LUTEICOLLIS.

Lygæus luteicollis, *Panz. Faun. Germ.* 93, 18—Halticus ochrocephalus, *Fieb. Weit. Beit.* i. 115, pl. 2, f. 4—Capsus propinquus, *H.-Sch. Wanz. Ins.* vi. 47, pl. 196, f. 606. *Kirschb. Caps.* 100—Halticus luteicollis, *Fieb. Eur. Hem.* 281—Halticocoris luteicollis, *Dougl. and Scott, Hem.* 480.

Europe.

2. LEPTOMEROCORIS ERYTHROCEPHALUS.

Capsus erythrocephalus, *H.-Sch. Nom.* 53. *Kirschb. Caps.* 102—Halticus erythrocephalus, *Fieb. Eur. Hem.* 281—Cimex saltator? *Rossi, Faun. Etr. Mant. Sp.* 513.

Germany.

3. LEPTOMEROCORIS PALLICORNIS.

Cicada aptera, *Linn. Faun. Suec.* 894—Acanthia pallicornis, *Fabr. Ent. Syst.* iv. 69. *Wolff, Icon. Cim.* 128, 122, pl. 13, f. 122—Salda pallicornis, *Fabr. Syst. Rhyn.* 115—Lygæus pallicornis, *Fall. Mon. Cim.* 95 — Astemma apterum, *Serv. Hist. Hem.* 284—Phytocoris pallicornis, *Fall. Hem. Suec.* i. 113—Attus arenarius, *Hahn. Wanz. Ins.* iii. 34, pl. 84, f. 255. · *Kirschb. Caps.* 101. *Guér. Icon.* pl. 56, f. 6—Halticus pallicornis, *Hahn. Wanz. Ins.* i. 114, pl. 18, f. 61. *Burm. Handb. Ent.* ii. 278. *Fieb. Eur. Hem.* 287—Capsus pallicornis, *Mey. Caps.* 110, 103. *Sahlb. Geoc. Fen.* 118. *Kirschb. Caps.* 102—Eyrecephala pallicornis, *Kol. Mel. Ent.* ii. 130—Capsus pallidicornis, *Flor, Rhyn. Liv.* i. 583—Halticus pallidicornis, *Fieb. Wien. Ent. Mon.* viii. 221.

a—c. Europe.

4. LEPTOMEROCORIS MACROCEPHALUS.

Halticus macrocephalus, *Fieb. Crit. Gen.* 47, sp. 12; *Eur. Hem.* 282.

South France. Corsica.

5. LEPTOMEROCORIS PUNCTICOLLIS.

Halticus puncticollis, *Fieb. Verh. Zool. Bot. Ges. Wien.* xx. 261.

Montenegro.

6. LEPTOMEROCORIS INTRICATUS.

Halticus intricatus, *Fieb. Wien. Ent. Mon.* viii. 220.
South Germany.

Div. 2.

Byrsoptera, *Spin. Ess. Hem.* 191. *Dougl. and Scott, Hem.* 351—
Malthacus, *Fieb. Crit. Gen.* 77, 80; *Eur. Hem.* 74, 312.

7. LEPTOMEROCORIS CARICIS.

Capsus Caricis, *Fall. Hem. Suec.* i. 123. *Mey. Caps.* 66. *Sahlb. Geoc.*
Fen. 92—Cyllecoris Caricis, *Hahn. Wanz. Ins.* ii. 100, pl. 60, f. 184
—Capsus rufifrons, *Fall. Cim. Suec.* 105. *Mey. Caps.* 105. *H.-Sch.*
Wanz. Ins. iii. 110, pl. 108, f. 338. *Kirschb. Caps.* 70, 114. *Flor,*
Rhyn. Liv. i. 622—Bryocoris rufifrons, *Sahlb. Geoc. Fen.* 124—
Halticus rufifrons, *Burm. Handb. Ent.* ii. 278—Byrsoptera erythro-
cephala, *Spin. Hem.* 191—Malthacus Caricis, *Fieb. Eur. Hem.* 313—
Byrsoptera Caricis, *Dougl. and Scott, Hem.* 352.

a—l. England. From Mr. Stephens' collection.
m, n. South France. Presented by F. Walker, Esq.
o, p. Europe. From Mr. Children's collection.

Div. 3.

Mecomma, *Fieb. Crit. Gen.* 60, pl. 6, f. 17; *Eur. Hem.* 69, 284—Sphyracephalus, *Dougl. and Scott, Hem.* 348.

8. LEPTOMEROCORIS AMBULANS.

Capsus ambulans, *Fall. Hem. Suec.* i. 126. *Hahn. Wanz. Ins.* iii. 109,
pl. 108, f. 335—337. *Mey. Caps.* 86, 67. *Zett. Ins. Lapp.* 279.
Sahlb. Geoc. Fen. 94. *Kirschb. Caps.* 76. *Flor, Rhyn. Liv.* i. 577—
Chlamydatus ochripes, *Curt. Brit. Ent.* xv. 693 — Mecomma
ambulans, *Fieb. Eur. Hem.* 284—Sphyracephalus ambulans, *Dougl.*
and Scott, Hem. 349.

Europe.

Div. 4.

Cyrtorhinus, *Fieb. Crit. Gen.* 51; *Eur. Hem.* 69, 284—Sphyrocephalus, p.,
Dougl. and Scott.

9. LEPTOMEROCORIS ELEGANTULUS.

Capsus elegantulus, *Mey. Caps.* 69, pl. 5, f. 2—Cyrtorhinus elegantulus,
Fieb. Eur. Hem. 285—Sphyracephalus elegantulus, *Dougl. and Scott,*
Hem. 351.

Europe.

Div. 5.

Tytthus, *Fieb. Wien. Ent. Mon.* viii. 82, pl. 2, f. 10.

10. LEPTOMEROCORIS PYGMÆUS.

Capsus pygmæus, *Zett. Ins. Lapp.* 279. *Flor, Rhyn. Liv.* i. 605—Tytthus pygmæus, *Fieb. Wien. Ent. Mon.* viii. 82.

Lapland. Livonia.

11. LEPTOMEROCORIS GEMINUS.

Capsus geminus, *Flor, Rhyn. Liv.* i. 606—Tytthus geminus, *Fieb. Wien. Ent. Mon.* viii. 82.

Livonia.

12. LEPTOMEROCORIS INSIGNIS.

Tytthus insignis, *Dougl. and Scott, Ent. M. Mag.* ii. 247.

England.

Div. 6.

Xenocoris, *Fieb. Crit. Gen.* 56; *Eur. Hem.* 71, 288.

13. LEPTOMEROCORIS VENUSTUS.

Xenocoris venustus, *Fieb. Mey. Crit. Gen.* 56, sp. 14.

Corsica. South Spain.

Div. 7.

Orthotylus, *Fieb. Crit. Gen.* 57; *Eur. Hem.* 71, 288—Litosoma, p. *Dougl. and Scott, Hem.*

14. LEPTOMEROCORIS ATRICAPILLUS.

Litosoma atricapilla, *Scott, Ent. M. Mag.* viii. 194.

Corsica.

15. LEPTOMEROCORIS ANGUSTUS.

Capsus angustus, *H.-Sch. Nom. Ent.* i. 49. *Mey. Caps.* i. 56, 19, pl. 2, f. 3. *Kirschb. Caps.* 77—Orthotylus angustus, *Fieb. Eur. Hem.* 288 —Litosoma angustus, *Dougl. and Scott, Hem.* 343.

Germany. Switzerland.

16. LEPTOMEROCORIS FLAVOSPARSUS.

Capsus flavosparsus, *Sahlb. Geoc. Fen.* 103. *Kirschb. Caps.* 89. *Flor, Rhyn. Liv.* i. 582—Phytocoris flavosparsus, *Boh. Vet. Ak. Forh.* 1852, 13, 18—Orthotylus flavosparsus, *Fieb. Eur. Hem.* 288—Litosoma flavosparsus, *Dougl. and Scott, Hem.* 341.

Europe.

17. LEPTOMEROCORIS FLAVINERVIS.

Capsus flavinervis, *Kirschb. Caps.* 79, 147—Orthotylus flavinervis, *Fieb. Eur. Hem.* 289—Litosoma flavinervis, *Dougl. and Scott, Hem.* 338.

Germany. Switzerland.

18. LEPTOMEROCORIS OBSOLETUS.

Orthotylus obsoletus, *Pict. Mey. Fieb. Eur. Hem.* 289.

Spain.

19. LEPTOMEROCORIS CONCOLOR.

Capsus concolor, *Kirschb. Caps.* 89, 155—Orthotylus concolor, *Fieb. Eur. Hem.* 289—Litosoma concolor, *Dougl. and Scott, Hem.* 340.

Germany.

20. LEPTOMEROCORIS NASSATUS.

Cimex nassatus, *Fabr. Mant. Ins.* ii. 304—Lygæus nassatus, *Fabr. Ent. Syst.* iv. 174; *Syst. Rhyn.* 236—Phytocoris nassatus, *Fall. Hem. Suec.* i. 80. *Kol. Mel. Ent.* ii. 117. *Zett. Ins. Lapp.* 272—Capsus nassatus, *Mey. Caps.* 50, 8. *Sahlb. Geoc. Fen.* 102. *Kirschb. Caps.* 78. *Flor, Rhyn. Liv.* i. 618—Lygus nassatus, *Hahn. Wanz. Ins.* i. 153, pl. 24, f. 78—Lygus icterocephalus, *Hahn. Wanz. Ins.* i. 149, pl. f. 75—Orthotylus nassatus, *Fieb. Eur. Hem.* 289—Litosoma nassatus, *Dougl. and Scott, Hem.* 337.

Siberia.

a, b. Europe. From Mr. Children's collection.

21. LEPTOMEROCORIS STRIICORNIS.

Capsus striicornis, *Kirschb. Caps.* 78, 143. *Flor, Rhyn. Liv.* i. 615—Orthotylus striicornis, *Fieb. Eur. Hem.* 289—Litosoma striicornis, *Dougl. and Scott, Hem.* 336.

Europe.

22. LEPTOMEROCORIS VIRIDINERVIS.

Capsus viridinervis, *Kirschb. Caps.* 78, 1142—Lygus floralis, *Hahn. Wanz. Ins.* i. 157, pl. 24, f. 81—Orthotylus viridinervis, *Fieb. Eur. Hem.* 290—Litosoma viridinervis, *Dougl. and Scott, Hem.* 335.

a, b. Europe.

23. LEPTOMEROCORIS DIAPHANUS.

Capsus diaphanus, *Kirschb. Caps.* 78—Orthotylus diaphanus, *Fieb. Eur. Hem.* 290—Litosoma diaphana, *Dougl. and Scott, Ent. M. Mag.* iv. 46.

England. Germany.

24. LEPTOMEROCORIS PELLUCIDUS.

Orthotylus pellucidus, *Garb. Bull. Soc. Ent. Ital.* i. 190.
Turin.

25. LEPTOMEROCORIS PALLIDUS.

Orthytylus pallidus, *Mey. Dür. Mitth. Schweiz. Ent. Ges.* iii. 209.
Switzerland.

26. LEPTOMEROCORIS OCHROTRICHUS.

Orthotylus ochrotrichus, *Fieb. Wien. Ent. Mon.* viii. 330.
England.

27. LEPTOMEROCORIS FIEBERI.

Orthotylus Fieberi, *Frei-Gessner, Mittheil. Schw. Ent. Ges.* 1864, 260.
Sarepta.

Div. 8.

Stenoparia, *Fieb. Verh. Zool. Bot. Ges. Wien.* xx. 255, pl. 6, f. 12.

28. LEPTOMEROCORIS PUTORI.

Stenoparia Putoni, *Fieb. Verh. Zool. Bot. Ges. Wien.* xx. 256.
Spain.

Div. 9.

Microsynamma, *Fieb. Wien. Ent. Mon.* viii. 74, pl. 1, f. 6.

29. LEPTOMEROCORIS SCOTTI.

Microsynamma Scotti, *Fieb. Wien. Ent. Mon.* viii. 75. *Scott. Ent. Ann.*
1864, 160, l. 5.
England.

Div. 10.

Stiphrosoma, *Fieb. Crit. Gen.* pl. 6, f. 12; *Eur. Hem.* 69, 280. *Dougl.
and Scott, Hem.* 481—Attus, *Hahn. Burm. Handb. Ent.* ii.—
Stiphrosomidæ, *Dougl. and Scott, Hem.* 35.

30. LEPTOMEROCORIS LEUCOCEPHALUS.

Cimex leucocephalus, *Linn. Syst. Nat.* 273; *Faun. Suec.* 940. *Deg. Ins.*
iii. 290. *Fabr. Mant. Ins.* ii. 304—Lygæus leucocephalus, *Fabr. Syst.
Rhyn.* 237. *Panz. Faun. Germ.* 92, 12. *Wolff, Icon. Cim.* 76, pl. 8,
f. 73—Capsus leucocephalus, *Mey. Caps.* 109, 100. *Sahlb. Geoo. Fen.*
117. *Flor, Rhyn. Liv.* i. 558. *Kirschb. Caps.* 86—Phytocoris leucocephalus, *Zett. Ins. Lapp.* 29. *Fall. Hem. Suec.* i. 111. *Hahn. Wanz.
Ins.* ii. 88, pl. 57, f. 174—Attus leucocephalus, *Burm. Handb. Ent.* ii.

276—Stiphrosoma leucocephala, *Fieb. Eur. Hem.* 281. *Dougl. and Scott, Hem.* 482—Cimex decrepitus, *Fabr. Ent. Syst.* iv. 125—Miris decrepitus, *Fabr. Syst. Rhyn.* 254.

Europe. Siberia.

31. LEPTOMEROCORIS NIGERRIMUS.

Capsus nigerrimus, *H.-Sch. Wanz. Ins.* iii. 87, pl. 101, f. 311—Stiphrosoma nigerrima, *Fieb. Eur. Hem.* 392.

Germany.

32. LEPTOMEROCORIS LURIDUS.

Phytocoris luridus, *Fall. Hem. Suec.* 112—Capsus luridus, *Hahn. Wanz. Ins.* ii. 87, pl. 101, f. 312—Stiphrosoma lurida, *Fieb. Eur. Hem.* 281.

Europe.

33. LEPTOMEROCORIS LIVIDUS.

Stiphrosoma livida, *Fieb. Crit. Gen.* 46, sp. 11; *Eur. Hem.* 281—Capsus obesus? *Muls. Ann. Soc. Linn.* iv. 165.

France. Corsica.

34. LEPTOMEROCORIS AMABILIS.

Stiphrosoma amabilis, *Dougl. and Scott, Ent. M. Mag.* v. 136.

Hebron.

35. LEPTOMEROCORIS ATROCÆRULEA.

Stiphrosoma atrocærulea, *Fieb. Wien. Ent. Mon.* viii. 329.

South Europe.

Div .11.

Heterotoma, *Latr. Fam. Nat.* 422. *Serv. Gen.* 236. *Fieb. Crit. Gen.* 58; *Eur. Hem.* 71, 290. *Dougl. and Scott, Hem.* 437.

36. LEPTOMEROCORIS MERIOPTERUS.

Cimex meriopterus, *Scop. Ent. Carn.* 382. *Rossi, Faun. Etr.* 1344—Capsus spissicornis, *Fabr. Syst. Rhyn.* 246. *Panz. Faun. Germ.* 215. *Enc. Meth.* pl. 373, f. 27. *Faun. Fr.* pl. 6, f. 8—Lygæus spissicornis, *Panz. Faun. Germ.* 2, 16—Heterotoma spissicornis, *Burm. Handb. Ent.* ii. 276. *Serv. Hem.* 283—Heterotoma meriopterus, *Fieb. Eur. Hem.* 290—Heterotoma merioptera, *Dougl. and Scott, Hem.* 438.

a—f. England. From Mr. Stephens' collection.
g—i. England. Presented by F. Walker, Esq.

Div. 12.

Heterocorydylus, *Fieb. Crit. Gen.* 59, pl. 6, f. 6; *Eur. Hem.* 71, 290. *Dougl. and Scott, Hem.* 432.

37. LEPTOMEROCORIS TUMIDICORNIS.

Capsus tumidicornis, *H.-Sch. Wanz. Ins.* iii. 84, pl. 100, f. 307. *Kirschb. Caps.* 84—Capsus Mali? *Boh. Nya Suec. Sp.* 20—Heterocordylus tumidicornis, *Fieb. Eur. Hem.* 291.
Germany.

38. LEPTOMEROCORIS TIBIALIS.

Capsus tibialis, *Hahn. Wanz. Ins.* i. 128, pl. 20, f. 66. *Kirschb. Caps.* 85 —Heterocordylus tibialis, *Dougl. and Scott, Hem.* 434.
Europe.

39. LEPTOMEROCORIS LEPTOCERUS.

Capsus leptocerus, *Kirschb. Caps.* 85—Capsus mutabilis? *Hahn. Wanz. Ins.* ii. 95, pl. 59, f. 180—Heterocordylus leptocorus, *Fieb. Eur. Hem.* 291.
Germany.

40. LEPTOMEROCORIS UNICOLOR.

Capsus unicolor, *Hahn. Wanz. Ins.* ii. 94, pl. 59, f. 179. *Kirschb. Caps.* 85—Heterotoma pulverulenta, *Klug. Burm. Handb. Ent.* ii. 275—Heterocordylus unicolor, *Fieb. Eur. Hem.* 291. *Dougl. and Scott, Hem.* 432.
a—f. England. Presented by F. Walker, Esq.

41. LEPTOMEROCORIS OBLONGUS.

Heterotoma oblonga, *Kol. Mel. Sp.* 110—Heterocordylus unicolor, *Fieb. Eur. Hem.* 392.
Germany.

42. LEPTOMEROCORIS PLANICORNIS.

Heterotoma planicornis, *H.-Sch. Wanz. Ins.* iii. 84, pl. 100, f. 306.
Germany.

Div. 13.

Chlamydatus, *Curt. Brit. Ent.* 693—Pachytoma, *Costa, A. S. E. F.* x. 289—Orthocephalus, *Fieb. Crit. Gen.* 60, pl. 6, f. 16. *Eur. Hem.* 71, 291. *Dougl. and Scott, Hem.* 429.

43. LEPTOMEROCORIS TRISTIS.

Orthocephalus tristis, *Pict. Mey. Fieb. Eur. Hem.* 292.
South Spain.

44. LEPTOMEROCORIS SIGNATUS.

Orthocephalus signatus, *Pict. Mey. Fieb. Eur. Hem.* 292.
South Spain.

45. LEPTOMEROCORIS SCHMIDTI.

Orthocephalus Schmidti, *Fieb. Crit. Gen.* 60, sp. 15 ; *Eur. Hem.* 292.
South Germany.

46. LEPTOMEROCORIS NEBULOSUS.

Orthocephalus nebulosus, *Pict. Mey. Fieb. Eur. Hem.* 293.
Spain.

47. LEPTOMEROCORIS VITTIPENNIS.

Capsus vittipennis, *H.-Sch. Wanz. Ins.* iii. 83, pl. 100, f. 305. *Sahlb.*
 Geoc. Fen. 120—Orthocephalus vittipennis, *Fieb. Eur. Hem.* 293.
Germany. Finland.

48. LEPTOMEROCORIS SALTATOR.

Capsus saltator, *Hahn. Wanz. Ins.* iii. 11, pl. 76, f. 236. *Mey. Caps.* 112,
 106. *Kirschb. Caps.* 83, 118—Capsus hirtus, *Curt. Brit. Ent.* xv. pl.
 693—Capsus mutabilis, *Flor, Rhyn. Liv.* i. 567—Orthocephalus
 mutabilis, *Fieb. Eur. Hem.* 293. *Dougl. and Scott, Hem.* 431.
a, b. England.

49. LEPTOMEROCORIS NITIDUS.

Capsus nitidus, *Mey. Rhyn.* sp. 107, pl. 6, f. 4—Orthocephalus nitidus,
 Fieb. Eur. Hem. 293.
Switzerland. Bohemia.

50. LEPTOMEROCORIS MUTABILIS.

Capsus mutabilis, *Fall. Mon. Cim.* 98, 4 ; *Hem. Suec.* i. 118. *Kirschb.*
 Caps. 83—Capsus pilosus, *Hahn. Wanz. Ins.* ii. 96, pl. 59, f. 181.
 Mey. Caps. 59, 24. *Flor. Rhyn. Liv.* i. 564—Orthocephalus mutabilis,
 Fieb. Eur. Hem. 293. *Dougl. and Scott, Hem.* 430.
Europe.

51. LEPTOMEROCORIS CORIACEUS.

Acanthia coriacea, *Fabr. Ent. Syst.* iv. 69—Salda coriacea, *Fabr. Syst. Rhyn.* 115—Orthocephalus mutabilis, *Fieb. Eur. Hem.* 293—Ortho-cephalus coriaceus, *Stal, Hem. Fabr.* i. 88.
Europe.

52. LEPTOMEROCORIS BREVIS.

Capsus brevis, *Panz. Faun. Germ.* 598. *Kirschb. Rhyn.* Sp. 109. *Mey. Rhyn.* Sp. 108—Orthocephalus Panzeri, *Fieb. Eur. Hem.* 294.
Germany. Switzerland.

53. LEPTOMEROCORIS MINOR.

Pachytoma minor, *Costa, A. S. E. F.* 289, pl. 6, f. 4—Chlomydatus minor, *Serv. Hem.* 285—Orthocephalus minor, *Fieb. Eur. Hem.* 294.
Europe.

54. LEPTOMEROCORIS BIVITTATUS.

Orthocephalus bivittatus, *Fieb. Wien. Ent. Mon.* viii. 221.
Sarepta.

55. LEPTOMEROCORIS RHYPAROPUS.

Orthocephalus rhyparopus, *Fieb. Wien. Ent. Mon.* viii. 222.
Sarepta.

56. LEPTOMEROCORIS FREYI.

Orthocephalus Freyi, *Fieb. Wien. Ent. Mon.* viii. 223.
Sarepta.

Div. 14.

Labops, *Burm. Handb. Ent.* ii. 279. *Fieb. Crit. Gen.* 61; *Eur. Hem.* 71, 294.

57. LEPTOMEROCORIS SAHLBERGI.

Capsus Sahlbergi, *Fall. Hem. Suec.* 116. *Sahlb. Geoc. Fen.* 118—Opthal-mocoris, *Zett. Ins. Lapp.* 280—Labops diopsis, *Burm. Handb. Ent.* ii. 279—Capsus diopsis, *H.-Sch. Wanz. Ins.* ix. 166, pl. 313, f. 961, 962—Labops Sahlbergi, *Fieb. Eur. Hem.* 294.
Europe.

58. LEPTOMEROCORIS BURMEISTERI.

Labops Burmeisteri, *Stal, Stett. Ent. Zeit.* xix. 189.
Kamtschatka.

Div. 15.

Diplacus, *Stal, Stett. Ent. Zeit.* xix. 183—Myrmecophyes, *Fieb. Verh. Zool. Bot. Ges. Wien.* xx. 253, pl. 6, f. 10.

59. LEPTOMEROCORIS OSCHANNINI.

Myrmecophyes Oschannini, *Fieb. Verh. Zool. Bot. Ges. Wien.* xx. 253. Russia.   •

60. LEPTOMEROCORIS ALBO-ORNATUS.

Diplacus albo-ornatus, *Stal, Stett. Ent. Zeit.* xix. 183, pl. 1, f. 3.
Siberia.

Div. 16.

Harpocera, *Curt. Brit. Ent.* xv.   *Fieb. Crit. Gen.* 63, pl. 6, f. 41; *Eur Hem.* 72, 296. *Dougl. and Scott, Hem.* 468.

61. LEPTOMEROCORIS THORACICUS.

Phytocoris thoracicus, *Fall. Mon. Cim.* 81; *Hem. Suec.* vii.—Capsus thoracicus, *Kirschb. Caps.* 73—Harpocera Burmeisteri, *Curt. Brit. Ent.* xv. pl. 709—Capsus antennatus, *Muls. Ann. Soc. Lin.* 1852, 130 —Capsus curvipes, *Mey. Caps.* 98, 86, pl. 5, f. 3—Capsus thoracicus, *Mey. Caps.* 102, 90, pl. 6, f. 5. *Kirschb. Caps.* 73, 82—Harpocera thoracica, *Fieb. Eur. Hem.* 297. *Dougl. ond Scott, Hem.* 469.

a—k. England.   From Mr. Stephens' collection.
l—t. England.   Presented by F. Walker, Esq.

Div. 17.

Megalodactylus, *Fieb. Crit. Gen.* 64 ; *Eur. Hem.* 73, 297.

62. LEPTOMEROCORIS MACULA-RUBRA.

Capsus macula-rubra, *Muls. Ann. Soc. Lin.* 1852, 138—Megalodactylus macula-rubra, *Fieb. Eur. Hem.* 297.
South France.

Div. 18.

Anoterops, *Fieb. Crit. Gen.* 65 ; *Eur. Hem.* 72, 297, 392. *Dougl. and Scott, Hem.* 384.

63. LEPTOMEROCORIS SETULOSUS.

Capsus setulosus, *Mey. Caps.* 53, 13, pl. 2, f. 1—Anoterops setulosus, *Fieb. Eur. Hem.* 298. *Dougl. and Scott, Hem.* 385.
England.   Switzerland.

Div. 19.

Cylindromelus, *Fieb. Eur. Hem.* 393.

64. LEPTOMEROCORIS SETULOSUS (bis lectum).

Capsus setulosus, *H.-Sch. Wanz. Ins.* iv. 30, pl. 120, f. 380—Cylindromelus setulosus, *Fieb. Eur. Hem.* 393.

Germany. Hungary.

Div. 20.

Oncotylus, *Fieb. Crit. Gen.* 66, pl. 6, f. 7; *Eur. Hem.* 72, 298, 393. *Dougl. and Scott, Hem.* 392.

65. LEPTOMEROCORIS DECOLOR.

Capsus decolor, *Fall. Hem. Suec.* i. 123. *Mey. Caps.* 86, 68. *Sahlb. Geoc. Fen.* 95. *Flor, Rhyn. Liv.* i. 555. *Kirschb. Caps.* 77—Lopus Chrysanthemi, *Hahn, Wanz. Ins.* i. 10, pl. 1, f. 4—Oncotylus decolor, *Fieb. Eur. Hem.* 298. *Dougl. and Scott, Hem.* 393.

Europe.

66. LEPTOMEROCORIS FENESTRATUS.

Oncotylus fenestratus, *Fieb. Crit. Gen.* 66, Sp. 19. *Eur. Hem.* 298.

Bohemia. Galicia.

67. LEPTOMEROCORIS TANACETI.

Phytocoris Tanaceti, *Fall. Hem. Suec.* i. 83—Miris Tanaceti, *Germ. Faun. Ins. Eur.* 16, 15—Capsus Tanaceti, *H.-Sch. Wanz. Ins.* iii. 851, pl. 101, f. 309. *Kirschb. Caps.* 80. *Flor, Rhyn. Liv.* i. 610—Oncotylus Tanaceti, *Fieb. Eur. Hem.* 299. *Dougl. and Scott, Hem.* 304.

a. Europe. From Mr. Children's collection.

68. LEPTOMEROCORIS PILOSUS.

Oncotylus pilosus, *Dougl. and Scott, Hem.* 395.

England.

69. LEPTOMEROCORIS HIPPOPHAES.

Capsus Hippophaes, *Mey. Cat.*—Oncotylus Hippophaes, *Fieb. Eur. Hem.* 299.

Vallais. South France.

70. LEPTOMEROCORIS PUNCTIPENNIS.

Oncotylus punctipennis, *Fieb. Wien. Ent. Mon.* viii. 225.
Sarepta.

Div. 21.

Conostethus, *Fieb. Crit. Gen.* 67; *Eur. Hem.* 72, 299, 393. *Dougl. and Scott, Hem.* 397.

71. LEPTOMEROCORIS ROSEUS.

Capsus roseus, *Fall, Hem. Suec.* i. 124—Capsus aridellus, *Flor, Rhyn. Liv.* i. 556—Conostethus roseus, *Fieb. Eur. Hem.* 299, 393. *Dougl. and Scott, Hem.* 398.

a. Europe. From Mr. Children's collection.

Div. 22.

Auchenocrepis, *Fieb. Crit. Gen.* 78; *Eur. Hem.* 74, 313.

72. LEPTOMEROCORIS FORELI.

Capsus Foreli, *Muls. Ann. Soc. Lin.* 1856, 130—Auchenocrepis dorsalis, *Fieb. Crit. Gen.* 78, Sp. 32—Auchenocrepis Foreli, *Fieb. Eur. Hem.* 313.
South Europe.

Div. 23.

Gnostus, *Fieb. Crit. Gen.* 82; *Eur. Hem.* 75—Teratoscopus, *Fieb. Eur. Hem.* 315.

73. LEPTOMEROCORIS PLAGIATUS.

Capsus plagiatus, *H.-Sch. Nom.* 50; *Wanz. Ins.* vi. 30, pl. 191, f. 587. *Panz. Faun. Germ.* 135, 10—Phytocoris institutus, *Fieb. Weit. Beit.* i. 104, pl. 2, f. 3—Teratoscopus plagiatus, *Fieb. Eur. Hem.* 316.
Germany. Switzerland.

Div. 24.

Hoplomachus, *Fieb. Crit. Gen.* 83; *Eur. Hem.* 75, 316. *Dougl. and Scott, Hem.* 395.

74. LEPTOMEROCORIS THUNBERGI.

Miris Thunbergii, *Germ. Faun. Ins. Eur.* 13, pl. 19—Phytocoris Thunbergii, *Fall. Hem. Suec.* i. 105—Capsus Thunbergii, *Mey. Caps.* pl. 59. *Sahlb. Geoc. Fen.* 110. *Kirsch. Caps.* 82. *Flor, Rhyn. Liv.* i. 608—Lopus Hieracii, *Hahn, Wanz. Ins.* i. 144, pl. 22, f. 73—Hoplomachus Thunbergi, *Fieb. Eur. Hem.* 316. *Dougl. and Scott, Hem.* 396.

a—f. Europe. From Mr. Children's collection.

75. LEPTOMEROCORIS BILINEATUS.

Capsus bilineatus, *Fall. Hem. Suec.* 122. *Hahn, Wanz. Ins.* iii. 70, pl. 94, bis f. 285. *Kirschb. Caps.* 82—Hoplomachus bilineatus, *Fieb. Eur. Hem.* 316.

Div. 25.

Pachyxiphus, *Fieb. Crit. Gen.* 84; *Eur. Hem.* 75, 316.

76. LEPTOMEROCORIS LINEELLUS.

Capsus lineellus, *Muls. Ann. Soc. Lin.* 1852, 113—Capsus croceipes, *Costa, Cent.* 1852—Pachyxiphus lineellus, *Fieb. Eur. Hem.* 317.
South Europe.

Div. 26.

Placochilus, *Fieb. Crit. Gen.* 85; *Eur. Hem.* 75, 317.

77. LEPTOMEROGORIS SELADONICUS.

Phytocoris seladonicus, *Fall. Hem. Suec.* 82—Capsus seladonicus, *H.-Sch. Wanz. Ins.* vi. 33, pl. 191, f. 590. *Mey. Caps.* Sp. 55—Placochilus seladonicus, *Fieb. Eur. Hem.* 317.
a. Europe. From Mr. Children's collection.

78. LEPTOMEROCORIS SAREPTANUS.

Placochilus Sareptanus, *Frei-Gessner, Mittheil. Schw. Ent. Ges.* 1864, 262.
Sarepta.

Div. 27.

Macrotylus, *Fieb. Crit. Gen.* 86; *Eur. Hem.* 76, 317.

79. LEPTOMEROCORIS LUNIGER.

Macrotylus luniger, *Fieb. Crit. Gen.* 86, Sp. 34; *Eur. Hem.* 318.
Switzerland.

80. LEPTOMEROCORIS LUTESCENS.

Macrotylus lutescens, *Fieb. Verh. Zool. Bot. Ges. Wien.* xx. 262.
Spain.

81. LEPTOMEROCORIS NIGRICORNIS.

Macrotylus nigricornis, *Fieb. Wien. Ent. Mon.* viii. 331.
South Europe.

Div. 28.

Macrocoleus, *Fieb. Crit. Gen.* 88; *Eur. Hem.* 76, 319. *Dougl. and Scott, Hem.* 386.

82. LEPTOMEROCORIS BICOLOR.

Macrocoleus bicolor, *Pict. Mey. Fieb. Eur. Hem.* 319.

South Spain.

83. LEPTOMEROCORIS PAYKULI.

Phytocoris Paykuli, *Fall. Hem. Suec.* i. 106—Capsus maculipennis, *H.-Sch. Nom.* 50. *Mey. Caps.* 81, 60, pl. 5, f. 1. *Kirschb. Caps.* 84—Macrocoleus Paykuli, *Fieb. Eur. Hem.* 319. *Dougl. and Scott, Hem.* 388.

a. Europe. From Mr. Children's collection.

84. LEPTOMEROCORIS AURANTIACUS.

Macrocoleus aurantiacus, *Fieb. Crit. Gen.* 88, Sp. 38; *Eur. Hem.* 320.

Corsica.

85. LEPTOMEROCORIS AUREOLUS.

Macrocoleus aureolus, *Fieb. Eur. Hem.* 320.

Germany.

86. LEPTOMEROCORIS SORDIDUS.

Capsus sordidus, *Kirschb. Caps.* 87—Macrocoleus sordidus, *Fieb. Eur. Hem.* 320.

Germany.

87. LEPTOMEROCORIS EXSANGUIS.

Capsus exsanguis, *H.-Sch. Nom.* 50 ; *Kirschb. Caps.* 79—Macrocoleus exsanguis, *Fieb. Eur. Hem.* 320.

Germany.

88. LEPTOMEROCORIS MOLLICULUS.

Phytocoris molliculus, *Fall. Hem. Suec.* i. 82—Capsus molliculus, *H.-Sch. Wanz. Ins.* vi. 32, pl. 19, f. 589. *Mey. Caps.* 78, 54. *Sahlb. Geoc. Fen.* 103. *Kirschb. Caps.* 80, 101. *Flor, Rhyn. Liv.* i. 611— Capsus ochroleucus, *Kirschb. Caps.* 88, 107—Macrocoleus molliculus, *Fieb. Eur. Hem.* 326. *Dougl. and Scott, Hem.* 387.

a. Europe. From Mr. Children's collection.

89. LEPTOMEROCORIS SOLITARIUS.

Capsus solitarius, *Mey. Rhyn.* Sp. 62, pl. 5, f. 4—Capsus seladonicus, *Kirschb. Caps.* 81—Macrocoleus solitarius, *Fieb. Eur. Hem.* 321.

Europe.

90. Leptomerocoris elevatus.

Macrocoleus elevatus, *Fieb. Crit. Gen.* 88, Sp. 37 ; *Eur. Hem.* 321.'
South France. Corsica.

91. Leptomerocoris chrysotrichus.

Macrocoleus chrysotrichus, *Fieb. Wien. Ent. Mon.* viii. 332.
South Russia.

92. Leptomerocoris pictus.

Macrocoleus pictus, *Fieb. Wien. Ent. Mon.* viii. 333.
South Europe.

Div. 29.

Macrolophus, *Fieb. Crit. Gen.* 89, pl. 6, f. 25, 32 ; *Eur. Hem.* 76, 321.
Dougl. and Scott, Hem. 381.

93. Leptomerocoris glaucescens.

Macrolophus glaucescens, *Fieb. Crit. Gen.* 89, Sp. 39 ; *Eur. Hem.* 321.
Bohemia.

94. Leptomerocoris nubilus.

Capsus nubilus, *H.-Sch. Panz. Faun. Germ.* 135, 9. *Mey. Caps.* 73
—Macrolophus nubilus, *Fieb. Eur. Hem.* 322. *Dougl. and Scott,*
Hem. 382.
Bavaria. Switzerland.

95. Leptomerocoris costalis.

Macrolophus costalis, *Fieb. Crit. Gen.* 89, Sp. 41; *Eur. Hem.* 322.
South Europe.

Div. 30.

Odontoplatys, *Fieb. Grit. Gen.* 33 ; *Eur. Hem.* 76, 322, 395.

96. Leptomerocoris bidentulus.

Capsus bidentulus, *H.-Sch. Wanz. Ins.* vi. 96, pl. 212, f. 668—Odontoplatys bidentulus, *Fieb. Eur.'Hem.* 322.
South Europe.

Div. 31.

Malacocoris, *Fieb. Crit. Gen.* 91, pl. 6, f. 53 ; *Eur. Hem.* 76, 322. *Dougl.*
and Scott, Hem. 383.

97. LEPTOMEROCORIS CHLORIZANS.

Lygæus chlorizans, *Block. Panz. Faun. Germ.* 18, 21—Phytocoris chlorizans, *Fall. Hem. Suec.* i. 82—Capsus chlorizans, *Mey. Caps.* 76, 50, pl. 4, f. 4. *Sahlb. Geoc. Fen.* 98. *Kirschb. Caps.* 73. *Flor, Rhyn. Liv.* i. 551—Malacocoris chlorizans, *Fieb. Eur. Hem.* 323. *Dougl. and Scott, Hem.* 382.

a. England.

98. LEPTOMEROCORIS SMARAGDINUS.

Malacocoris smaragdinus, *Fieb. Crit. Gen.* 91, Sp. 42 ; *Eur. Hem.* 323. Bohemia.

99. LEPTOMEROCERIS ALBOPUNCTATUS.

Malacocoris albopunctatus, *Garb. Bull. Soc. Ent. Ital.* i. 194. Turin.

Div. 32.

Cyrtopeltis, *Fieb. Crit. Gen.* 92, pl. 6, f. 29 ; *Eur. Hem.* 76, 323.

100. LEPTOMEROCORIS GENICULATUS.

Cyrtopeltis geniculatus, *Pict. Mey. Fieb. Eur. Hem.* 325. South Spain.

Div.?

101. LEPTOMEROCORIS PROLIXUS.

prolixus, *Stal, Stett. Ent. Zeit.* xix. 187. Siberia.

102. LEPTOMEROCORIS GILVIPES.

gilvipes, *Stal, Stett. Ent. Zeit.* xix. 187. Sitka.

103. LEPTOMEROCORIS SERICANS.

sericans, *Stal, Stett. Ent. Zeit.* xix. 188. Sitka.

104. LEPTOMEROCORIS MUNDULUS.

mundulus, *Stal, Stett. Ent. Zeit.* xix. 188. Siberia.

South America.

105. LEPTOMEROCORIS CUNEATUS.

Capsus cuneatùs, *Stal, Rio Jan. Hem.* 55. Rio Janeiro.

Madeira.

106. LEPTOMEROCORIS OBESULUS.

Capsus? obesulus, *Wlln. Ann. Nat. Hist. 3rd Ser.* i. 124.

a—z. Madeira. From Mr. Wollaston's collection.

South Asia.

Div. n.

107. LEPTOMEROCORIS SIMPLEX.

Fuscescens, fusiformis, subtilissime punctatus, subtus testaceus; caput testaceum; oculi valde prominuli; antennæ gracillimæ, corpori æquilongæ, articulo 1o -capite longiore basi testaceo; prothorax testaceo marginatus et biplagiatus; scutelli apex pedesque testacei; ✳ *corii costa testaceo biplagiata; membrana cinerea.*

Brownish, fusiform, very finely punctured, testaceous beneath. Head testaceous, short. Eyes very prominent. Antennæ very slender, as long as the body; first joint testaceous towards the base, longer than the head; second much longer than the first and than the third; fourth much shorter than the third. Prothorax convex, with a testaceous border, and with a testaceous patch on each side of the disk; a distinct transverse furrow near the fore border. Scutellum testaceous at the tip. Legs testaceous, slender. Corium with two testaceous costal patches. Membrane cinereous. Length of the body 1¼ line.

a, b. Ceylon. Presented by Dr. Thwaites.

Eastern Isles.

Div. 1.

Ocypus, *Mtrz. A. S. E. F.* 4me *Sér.* i. 67—Coridromius, *Sgnt.*—Stiphrosoma? *Fieb.*

108. LEPTOMEROCORIS VARIEGATUS.

Ocypus variegatus, *Mtrz. A S. E. F.* 4me *Sér.* i. 68.

New Caledonia.

Australia.

109. LEPTOMEROCORIS ANTENNATUS.

Fœm. Ferrugineus, ellipticus, subtilissime punctatus, subtus flavus; oculi prominuli; rostrum coxas intermedias attingens; antennæ validæ, corpore paullo breviores, articulo 1o crasso capite paullo longiore, 2i dimidio apicali crasso basi flavo, 3o piceo basi flavo; prothoracis anguli postici prominuli, obtusi; propectus piceo sexlineatum; pedes flavi, femorum fasciis duabus latis tibiis apice tarsisque rufis; membrana cinerea, rufo venosa.

U

Female. Ferruginous, elliptical, very finely punctured, yellow beneath. Head triangular. Eyes piceous, prominent. Rostrum extending to the middle coxæ; tip black. Antennæ stout, a little shorter than the body; first joint thick, a little longer than the head; second about twice the length of the first, thick for more than half the length from the tip, yellow towards the base; third piceous, yellow towards the base, a little shorter than the first. Hind angles of the prothorax prominent, rounded, obtuse. Propectus on each side with three piceous lines, which are parallel to the sides of the prothorax. Legs yellow, slender, moderately long; femora with two broad red bands, rather stout; tibiæ at the tips and tarsi red. Membrane and hind wings cinereous, the former with red veins. Length of the body 3½ lines.

a. Australia. From Mr. Damel's collection.

New Zealand.

110. Leptomerocoris Maoricus.

Mas et fœm. *Testaceus aut fuscus, fusiformis, subtilissime punctatus; oculi subprominuli; rostrum coxas posticas attingens; antennæ corpori æquilongæ, articulo 1o capite paullo longiore, 2o apice 3o 4oque nigris; corium apud marginem exteriorem rufo uninotatum; membrana cinerea.*

Male and female. Testaceous, fusiform, very finely punctured, paler beneath, occasionally brown. Head short-triangular. Eyes piceous, slightly prominent. Rostrum extending to the hind coxæ; tip black. Antennæ as long as the body; first joint a little longer than the head; second about twice the length of the first, black at the tip; third and fourth black; third much shorter than the second; fourth shorter than the third. Legs slender. Corium with a red mark on the middle of the exterior border. Membrane cinereous. Length of the body 1½ line.

â—h. New Zealand. Presented by Capt. J. C. Ross.

Genus 7. EURYMEROCORIS.

Eurymerocoris, *Kirschb. Caps.* 31.

Div. 1.

Atractotomus, *Fieb. Crit. Gen.* 62, pl. 6, f. 39; *Eur. Hem.* 71, 294. *Dougl. and Scott, Hem.* 434.

1. Eurymerocoris sulcicornis.

Capsus sulcicornis, *Kirschb. Caps.* Sp. 125—Atractotomus sulcicornis, *Fieb. Eur. Hem.* 295.

Germany. Switzerland.

2. EURYMEROCORIS TIGRIPES.

Capsus magnicornis, *Hahn, Wanz. Ins.* i. 130, pl. 20, f. 67. *Mey. Rhyn.*
  Sp. 29, pl. 2, f. 4—Capsus tigripes, *Muls. Ann. Soc. Lin.* (1852), 129
  —Atractotomus tigripes, *Fieb. Eur. Hem.* 295.
Europe.

3. EURYMEROCORIS NIGRIPES.

Atractotomus nigripes, *Pict. Mey. Fieb. Eur. Hem.* 295.
South Spain.

4. EURYMEROCORIS FEMORALIS.

Atractotomus femoralis, *Fieb. Crit. Gen.* 62, Sp. 16; *Eur. Hem.* 295.
Bohemia.

5. EURYMEROCORIS RUFUS.

Atractotomus rufus, *Fieb. Crit. Gen.* 62, Sp. 17; *Eur. Hem.* 296.
Bohemia.

6. EURYMEROCORIS OCULATUS.

Capsus oculatus, *Kirschb. Caps.* 90—Atractotomus albipes, *Fieb. Crit.
  Gen.* 62, Sp. 18—Atractotomus oculatus, *Fieb. Eur. Hem.* 296.
Germany. Switzerland.

7. EURYMEROCORIS MALI.

Capsus planicornis, *H.-Sch. Wanz. Ins.* iii. 84, pl. 100, f. 306—Capsus
  Mali, *Mey. Rhyn.* Sp. 30, pl. 2, f. 5. *Kirschb. Caps.* 91—Atracto-
  tomus Mali, *Fieb. Eur. Hem.* 296.
Germany.

8. EURYMEROCORIS MAGNICORNIS.

Capsus magnicornis, *Fall. Mon. Cim.* 997; *Hem. Suec.* i. 119; *Zett. Ins.
  Lapp.* 278. *Kirschb. Caps.* 91. *Flor, Rhyn. Liv.* i. 575—Atractotomus magnicornis, *Fieb. Eur. Hem.* 296. *Dougl. and Scott, Hem.*
  435.
a—e. England. From Mr. Stephens' collection.
f—h. England. Presented by F. Walker, Esq.

9. EURYMEROCORIS RHODANI.

Atractotomus Rhodani, *Mey. Fieb. Eur. Hem.* 296.
Switzerland.

10. EURYMEROCORIS PUNCTIPES.

Atractotomus punctipes, *Fieb. Wien. Ent. Mon.* viii. 224.
Sarepta.

11. EURYMEROCORIS PINI.

Atractotomus Pini (*Dougl. and Scott*), *Fieb. Wien. Ent. Mon.* viii. 224.
England.

Div. 2.

Tinicephalus, *Fieb. Crit. Gen.* 68, pl. 6, f. 11; *Eur. Hem.* 73, 299. *Dougl. and Scott, Hem.* 390.

12. EURYMEROCORIS RUBIGINOSUS.

Tinicephalus rubiginosus, *Pict. Mey. Fieb. Eur. Hem.* 300.
South Spain.

13. EURYMEROCORIS HORTULANUS.

Capsus hortulanus, *Mey. Rhyn.* 77, pl. 7, f. 3. *Kirschb. Caps.* 89—Tinicephalus hortulanus, *Fieb. Eur. Hem.* 300.
Germany. Switzerland.

14. EURYMEROCORIS DISCREPANS.

Tinicephalus discrepans, *Fieb. Grit. Gen.* 68, Sp. 20; *Eur. Hem.* 300.
South France. Corsica.

15. EURYMEROCORIS OBSOLETUS.

Tinicephalus obsoletus, *Dougl. and Scott, Hem.* 391, pl. 13, f. 1. *Fieb. Wien. Ent. Mon.* viii. 226.
England.

Div. 3.

Tragiscus, *Fieb. Crit. Gen.* 69—Tragiscocoris, *Fieb. Eur. Hem.* 73, 300.

16. EURYMEROCORIS FIEBERI.

Tragiscus Ficheri, *Mey. Dür. Fieb. Crit. Gen.* 69, Sp. 21—Tragiscocoris Fieberi, *Fieb. Eur. Hem.* 301.

Div. 4.

Brachyarthrum, *Fieb. Crit. Gen.* 70; *Eur. Hem.* 73, 301.

17. EURYMEROCORIS LIMITATUS.

Brachyarthrum limitatum, *Fieb. Crit. Gen.* 70, Sp. 22; *Eur. Hem.* 301.
Bohemia.

18. EURYMEROCORIS PINETELLUS.

Phytocoris pinetella, *Zett. Ins. Lapp.* 276—Capsus Pinetellus, *Kirschb. Caps.* 76—Brachyarthrum Pinetellum, *Fieb. Eur. Hem.* 301—Phytocoris nigriceps, *Boh. K. V. Acad. Forh.* 1852, 15.
Europe.

Div. 5.

Criocoris, *Fieb. Crit. Gen.* 71 ; *Eur. Hem.* 73, 301.

19. EURYMEROCORIS CRASSICORNIS.

Phytocoris crassicornis, *Hahn, Wanz. Ins.* ii. 90, pl. 57, f. 176—Capsus crassicornis, *Wanz. Ins.* iii. 85, pl. 101, f. 308. *Kirschb. Caps.* 91— Criocoris crassicornis, *Fieb. Eur. Hem.* 302.

Europe.

20. EURYMEROCORIS NIGRIPES.

Criocoris nigripes, *Fieb. Eur. Hem.* 394.

Germany.

21. EURYMEROCORIS TIBIALIS.

Criocoris tibialis, *Fieb. Wien. Ent. Mon.* viii. 227.

South France.

Div. 6.

Liops, *Fieb. Verh. Zool. Bot. Ges. Wien.* xx. 254, pl. 6, f. 11.

22. EURYMEROCORIS PUNCTICOLLIS.

Liops puncticollis, *Fieb. Ver. Zool. Bot. Ges. Wien.* xx. 254.
Spain.

Div. 7.

Plagiognathus, *Fieb. Crit. Gen.* 72 ; *Eur. Hem.* 73, 302. *Dougl. and Scott, Hem.* 400.

23. EURYMEROCORIS ARBUSTORUM.

Capsus arbustorum, *Fabr. Ent. Syst.* iv. 175 ; *Syst. Rhyn.* 238. *Hahn, Wanz. Ins.* iii. 80, pl. 99, f. 300. *Mey. Caps.* 64, 33, pl. 3, f. 1. *Sahlb. Geoc. Fen.* 115, 55. *Kirschb. Caps.* 99, 145. *Flor, Rhyn. Liv.* i. 602—Phytocoris arbustorum, *Fall. Hem. Suec.* i. 104. *Zett. Ins. Lapp.* 275—Capsus brunnipennis, *Mey. Caps.* 66, 35, pl. 3, f. 3. *Kirschb. Caps.* 99, 144—Capsus hortensis, *Mey. Caps.* 65, 34, pl. 3, f. 2. *Kirschb. Caps.* 106—Plagiognathus arbustorum, *Fieb. Eur. Hem.* 302. *Dougl. and Scott, Hem.* 402—Phytocoris lugubris, *Hahn, Wanz. Ins.* ii. 138, pl. 72, f. 225.

a, b. England.

24. EURYMEROCORIS INFUSCATUS.

Plagiognathus infuscatus, *Pict. Mey. Fieb. Eur. Hem.* 303.
Spain.

25. EURYMEROCORIS FULVIPENNIS.

Capsus fulvipennis, *Kirschb. Caps.* 99—Plagiognathus fulvipennis, *Fieb. Eur. Hem.* 303.

Germany.   Switzerland.

26. EURYMEROCORIS VIRIDULUS.

Phytocoris viridulus, *Fall. Hem. Suec.* i. 105—Capsus viridulus, *Hahn, Wanz. Ins.* ii. 136, pl. 72, f. 221. *Mey. Caps.* 77, 51, pl. 7, f. 2. *Sahlb. Geoc. Fen.* 103. *Kirschb. Caps.* 98. *Flor, Rhyn. Liv.* i. 595. Plagiognathus viridulus, *Fieb. Eur. Hem.* 303. *Dougl. and Scott, Hem.* 302.

a. Europe.   From Mr. Children's collection.

27. EURYMEROCORIS BOHEMANNI.

Phytocoris Bohemanni, *Fall. Hem. Suec.* 106—Phytocoris ruficollis, *Fall. Hem. Suec.* 107—Capsus furcatus, *H.-Sch. Wanz. Ins.* iv. 79, pl. 132, f. 408, 409. *Kirschb. Caps.* 95—Plagiognathus Bohemanni, *Fieb. Eur. Hem.* 303—Neocoris Bohemanni, *Dougl. and Scott, Hem.* 423.

Europe.

28. EURYMEROCORIS SPILOTUS.

Plagiognathus spilotus, *Fieb. Crit. Gen.* 72, Sp. 23; *Eur. Hem.* 304.

Corsica.

Div. 8.

Apocremnus, *Fieb. Crit. Gen.* 73; *Eur. Hem.* 74, 304.   *Dougl. and Scott, Hem.* 403.

29. EURYMEROCORIS ANCORIFER.

Apocremnus ancorifer, *Fieb. Crit. Gen.* 73, Sp. 24; *Eur. Hem.* 304.

South France.   Spain.

30. EURYMEROCORIS AMBIGUUS.

Phytocoris ambiguus, *Fall. Mon. Cim.* 891. *Hem. Suec.* i. 99. *Zett. Ins. Lapp.* 274—Capsus ambiguus, *H.-Sch. Wanz. Ins.* vi. 43, pl. 195, f. 602. *Mey. Caps.* 60, 27. *Sahlb. Geoc. Fen.* 114, 51. *Kirschb. Caps.* 94. *Flor, Rhyn. Liv.* i. 627—Capsus obscurus, *Kirschb. Caps.* 92—Capsus Betulæ, *Kirschb. Caps.* 94—Apocremnus ambiguus, *Fieb. Eur. Hem.* 305. *Dougl. and Scott, Hem.* 404.

Europe.

31. EURYMEROCORIS QUERCUS.

Capsus Quercus, *Kirschb. Caps.* 93, 163—Apocremnus Quercus, *Fieb. Eur. Hem.* 305. *Dougl. and Scott, Hem.* 409.

Europe.

32. EURYMEROCORIS VARIABILIS.

Phytocoris variabilis, *Fall. Hem. Suec.* i. 98. *Hahn, Wanz. Ins.* ii. 137, pl. 72, f. 224. *Zett. Ins. Lapp.* 275—Capsus variabilis, *Mey. Caps.* 68, 38, pl. 3, f. 4. *Sahlb. Geoc. Fen.* 115. *Kirschb. Caps.* 23. *Flor, Rhyn. Liv.* i. 592—Capsus roseus, *H.-Sch. Wanz. Ins.* vi. 46, pl. 195, f 604—Apocremnus variabilis, *Fieb. Eur. Hem.* 305. *Dougl. and Scott, Hem.* 408.

Europe.

33. EURYMEROCORIS SIMILLIMUS.

Capsus simillimus, *Kirschb. Caps.* 93, 165—Apocremnus simillimus, *Fieb. Eur. Hem.* 305. *Dougl. and Scott, Hem.* 410.

Europe.

Div. 9.

Psallus, *Fieb. Crit. Gen.* 305; *Eur. Hem.* 74, 305. *Dougl.-and Scott, Hem.* 410.

34. EURYMEROCORIS SALICELLUS.

Capsus salicellus, *Mey. Caps.* 74, 47. *H.-Sch. Wanz. Ins.* vi. 47, pl. 196, f. 605. *Flor, Rhyn. Liv.* i. 590—Psallus salicellus, *Fieb. Eur. Hem.* 306. *Dougl. and Scott, Hem.* 411.

Europe.

35. EURYMEROCORIS QUERCETI.

Phytocoris Querceti, *Fall. Hem. Suec.* i. 102—Capsus sanguineus, *Kirschb. Caps.* 97—Psallus Querceti, *Fieb. Eur. Hem.* 306. *Dougl. and Scott, Hem.* 413.

Europe.

36. EURYMEROCORIS ALNI.

Lygæus Alni, *Fabr. Ent. Syst.* iv. 175; *Syst. Rhyn.* 238—Phytocoris Querceti, *Fall. Hem. Suec.* i. 102—Capsus sanguineus, *Kirschb. Caps.* 97—Psallus Querceti, *Fieb. Eur. Hem.* 306. *Dougl. and Scott, Hem.* 412—Psallus Alni, *Stal, Hem. Fabr.* i. 88. *Dougl. and Scott, Hem.* 414.

Europe.

37. EURYMEROCORIS SANGUINEUS.

Lygæus sanguineus, *Fabr. Ent. Syst.* iv. 175. *Syst. Rhyn.* 238—Phytocoris sanguineus, *Fall. Hem. Suec.* i. 102—Capsus sanguineus, *H.-Sch. Nom. Ent.* i. 51; *Wanz. Ins.* iii. 70. *Mey. Caps.* 75, 49. *Sahlb. Geoc. Fen.* 107. *Flor, Rhyn. Liv.* i. 588—Psallus sanguineus, *Fieb. Eur. Hem.* 306. *Dougl. and Scott, Hem.* 413.

Europe.

38. EURYMEROCORIS SCHOLTZI.

Capsus Scholtzi, *Mey. Caps.*—Pallus Scholtzi, *Fieb. Eur. Hem.* 306.

Switzerland.

39. EURYMEROCORIS ALBICINCTUS.

Capsus albicinctus, *Kirschb. Caps.* 96—Psallus albicinctus, *Fieb. Eur. Hem.* 307.

Germany. Switzerland.

40. EURYMEROCORIS SALIGIS.

Capsus Salicis, *Kirschb. Caps.* 97, 174—Psallus Salicis, *Fieb. Crit. Gen.* 74, Sp. 25; *Eur. Hem.* 307. *Dougl. and Scott, Hem.* 415.

Europe.

41. EURYMEROCORIS LEPIDUS.

Capsus variabilis, *var. Mey. Dür. Caps.*—Pallus lepidus, *Fieb. Wien. Ent. Mon.* ii. 337; *Eur. Hem.* 307. *Dougl. and Scott, Hem.* 416.

England. Switzerland.

42. EURYMEROCORIS NOTATUS.

Psallus notatus, *Fieb. Crit. Gen.* 74, Sp. 30; *Eur. Hem.* 307.

South France.

43. EURYMEROCORIS VITELLINUS.

Capsus vitellinus, *Scholtz, Arb. Ver.* 26—Psallus vitelliuus, *Fieb. Eur. Hem.* 307.

Germany. Switzerland.

44. EURYMEROCORIS DILUTUS.

Psallus dilutus, *Mey. Fieb. Wien. Ent. Mon.* ii. 338; *Eur. Hem.* 308. *Dougl. and Scott, Hem.* 417.

England. Switzerland.

45. EURYMEROCORIS ARGYROTRICHUS.

Psallus argyrotrichus, *Fieb. Eur. Hem.* 308.

Germany. Spain.

46. EURYMEROCORIS ROSEUS.

Lygæus roseus, *Fabr. Syst. Rhyn.* 238—Phytocoris roseus, *Fall. Hem. Suec.* i. 101. *Zett. Ins. Lapp.* 275—Capsus roseus, *H.-Sch. Nom. Ent.* i. 49; *Wanz. Ins.* iii. 71, pl. 96, f. 287. *Mey. Caps.* 67, 37. *Sahlb. Geoc. Fen.* 107. *Kirschb. Caps.* 96. *Flor, Rhyn. Liv.* i. 591 —Psallus roseus, *Fieb. Eur. Hem.* 308. *Dougl. and Scott, Hem.* 417.

Europe.

47. EURYMEROCORIS KIRSCHBAUMI.

Capsus roseus, *Kirschb. Caps.* 96—Psallus Kirschbaumi, *Fieb. Eur. Hem.* 308.

Germany.

48. EURYMEROCORIS VARIANS.

Capsus varians, *H.-Sch. Wanz. Ins.* vi. 45, pl. 195, f. 603. *Mey. Caps.* 39
Capsus decoloratus, *Muls. Ann. Soc. Linn.* 1852, 124—Psallus
insignis, *Fieb. Crit. Gen.* 74, Sp. 27—Psallus varians, *Fieþ. Eur.
Hem.* 309. *Dougl. and Scott, Hem.* 418.
Europe.

49. EURYMEROCORIS DISTINCTUS.

Psallus distinctus, *Fieb. Wien. Ent. Mon.* ii. 337; *Crit. Gen.* 74, Sp. 26;
Eur. Hem. 309. *Dougl. and Scott, Hem.* 419.
England. Bohemia. Switzerland.

50. EURYMEROCORIS DIMINUTUS.

Capsus diminutus, *Kirschb. Caps.* 96—Psallus diminutus, *Fieb. Eur.
Hem.* 309.
Germany. Switzerland.

51. EURYMEROCORIS FIEBERI.

Psallus Fieberi, *Dougl. and Scott, Hem.* 420. *Fieb. Wien. Ent. Mon.* viii.
227.
England.

52. EURYMEROCORIS ELEGANS.

Psallus elegans, *Jakowlew, Horæ Soc. Ent. Ross,* iv. 158.
Astrachan.

53. EURYMEROCORIS FUSCOVENOSUS.

Psallus? fuscovenosus, *Fieb. Wien. Ent. Mon.* viii. 330.
Sarepta.

54. EURYMEROCORIS WHITEI.

Psallus Whitei, *Dougl. and Scott, Ent. M. Mag.* v. 263.
Scotland.

55. EURYMEROCORIS CROTCHI.

Psallus Crotchi, *Scott, Ent. Zeit. Stett.* 1870, 99.
Spain.

Div. 10.

Sthenarus, *Fieb. Crit. Gen.* 75; *Eur. Hem.* 74, 309. *Dougl. and Scott,
Hem.* 421.

56. Eurymerocoris Roseri.

Capsus Roseri, *H.-Sch. Wanz. Ins.* iv. 78, pl. 132, f. 407. *Mey. Caps.* 94. *Kirschb. Caps.* 87—Sthenarus Roseri, *Fieb. Eur. Hem.* 309.
Germany. Switzerland.

57. Eurymerocoris Rotermundi.

Capsus Rotermundi, *Scholtz, Arb. Ver.* 1846, 42. *Flor, Rhyn. Liv.* i. 594 —Sthenarus Rotermundi, *Fieb. Eur. Hem.* 310. *Dougl. and Scott, Hem.* 422.
Europe.

58. Eurymerocoris vittatus.

Sthenarus vittatus, *Fieb. Crit. Gen.* 75, Sp. 31; *Eur. Hem.* 310.
Bohemia.

Div. 11.

Astemma, *Serv. Hist. Hem.* 284—Halticus, p., *Burm. Handb. Ent.* ii. 278 —Eurycephala, *Blanch. Hist. Nat. Ins.* iii. 141—Agalliastes, *Fieb. Crit. Gen.* 76; *Eur. Hem.* 74, 310. *Dougl. and Scott, Hem.* 426.

59. Eurymerocoris evanescens.

Pachystoma evanescens, *Boh.*—Capsus evanescens, *Kirschb. Caps.* 103— Agalliastes evanescens, *Fieb. Eur. Hem.* 310; *Wien. Ent. Mon.* viii. 229.
Europe.

60. Eurymerocoris albipennis.

Phytocoris albipennis, *Fall. Hem. Suec.* 107—Capsus albipennis, *Hahn, Wanz. Ins.* ii. 91, pl. 57, f. 177. *Kirschb. Caps.* 92—Agalliastes albipennis, *Fieb. Eur. Hem.* 311.
Europe.

61. Eurymerocoris saltitans.

Lygæus saltitans, *Fall. Mon. Cim.* 96—Phytocoris saltitans, *Fall. Hem. Suec.* i. 114—Capsus saltitans, *Kirschb. Caps.* Sp. 153—Astemma saltitans, *Serv. Hist. Hem.* 284—Halticus saltitans. *Burm. Handb. Ent.* ii. 278—Eurycephala saltitans, *Blanch. Hist. Nat. Ins.* iii. 141 —Chlamydatus marginatus, *Curt. Brit. Ent.* xv. 603—Capsus saltitans, *Sahlb. Geoc. Fen.* 119. *Kirschb. Caps.* 103. *Flor, Rhyn. Liv.* i. 603—Agalliastes saltitans, *Fieb. Eur. Hem.* 311. *Dougl. and Scott, Hem.* 428.

a—d. England. From Mr. Stephens' collection.

62. EURYMEROCORIS VERBASCI.

Capsus Verbasci, *H.-Sch. Mey. Caps.* 42, pl. 4, f. 1.   *Kirschb. Caps.* 98—
Agalliastes Verbasci, *Fieb. Eur. Hem.* 311.
Germany.   Switzerland.

63. EURYMEROCORIS PUNCTATUS.

Agalliastes punctatus, *Pict. Mey. Fieb. Eur. Hem.* 311.
South Spain.

64. EURYMEROCORIS ONUSTUS.

Agalliastes onustus, *Pict. Mey. Fieb. Eur. Hem.* 312.
Spain.

65. EURYMEROCORIS PULICARIUS.

Phytocoris pulicarius, *Fall. Mon. Cim.* 95; *Hem. Suec.* 113—Capsus
saliens, *Wolff, Icon. Cim.* 200, pl. 20, f. 194—Attus pulicarius, *Hahn,
Wanz. Ins.* i. 117, pl. 18, f. 62. *Burm. Handb. Ent.* ii. 227—Capsus
pulicarius, *Mey. Caps.* 110, 102. *Sahlb. Geoc. Fen.* 119. *Kirschb.
Caps.* 101. *Flor, Rhyn. Liv.* i. 600—Agalliastes pulicarius, *Fieb.
Eur. Hem.* 312; *Wien. Ent. Mon.* viii. 2. *Dougl. and Scott, Hem.*
427.
Europe.   Siberia.

66. EURYMEROCORIS MODESTUS.

Capsus modestus, *Mey. Rhyn.* Sp. 40, pl. 3, f. 5—Capsns atropurpureus,
Kirschb. Rhyn. 101—Capsus gracilicornis, *Scholtz, Arb. Ver.* 1846,
106—Agalliastes modestus, *Fieb. Eur. Hem.* 313; *Wien. Ent. Mon.*
viii. 231.
Europe.

67. EURYMEROCORIS VITTATUS.

Agalliastes vittatus, *Fieb. Eur. Hem.* 312.
Austria.

68. EURYMEROCORIS LUGUBRIS.

Agalliastes lugubris, *Fieb. Eur. Hem.* 312.
Prussia.

69. EURYMEROCORIS WILKINSONII.

Agalliastes Wilkinsonii, *Dougl. and Scott, Ent. M. Mag.* ii. 273.
England.

70. EURYMEROCORIS TIBIALIS.

Agalliastes tibialis, *Fieb. Wien. Ent. Mon.* viii. 228.
Sarepta.

71. EURYMEROCORIS MEYERI.

Agalliastes Meyeri, *Fieb. Wien. Ent. Mon.* viii. 231.
Switzerland.

72. EURYMEROCORIS PRASINUS.

Agalliastes prasinus, *Fieb. Wien. Ent. Mon.* viii. 228.
Sarepta.

73. EURYMEROCORIS ALUTACEUS.

Agalliastes alutaceus, *Fieb. Verh. Zool. Bot. Ges. Wien.* xx. 262.
Spain.

74. EURYMERŌCORIS ABSINTHII.

Agalliastes Absinthii, *Scott, Stett. Ent. Zeit.* 1870, 100.
Martigny.

75. EURYMEROCORIS PALLIPES.

Agalliastes pallipes, *Jakowlew, Horæ Soc. Ent. Ross.* iv. 150.
Chwalynsk.

76. EURYMEROCORIS KIRGISICUS.

Agalliastes Kirgisicus, *Becker, MS. Frei-Gessner, Mittheil. Schw. Ent. Ges.* 1864, 261.
Sarepta.

77. EURYMEROCORIS KOLENATII.

Capsus Kolenatii, *Flor, Rhyn. Liv.* i. 585—Agalliastes Kolenatii, *Fieb. Wien. Ent. Mon.* viii. 230.
Livonia.

78. EURYMEROCORIS NIGRITULUS.

Phytocoris nigritulus, *Zett. Faun. Lapp.* 271—Capsus nigritulus, *Flor, Rhyn. Liv.* i. 599—Agalliastes nigritulus, *Fieb. Wien. Mon.* viii. 230.
Sweden. Livonia.

Div. 12.

Pachylops, *Fieb. Crit. Gen.* 53, pl. 6, f. 26; *Eur. Hem.* 70, 285.

79. EURYMEROCORIS CHLOROPTERUS.

Capsus chloropterus, *Kirschb. Caps.* 89—Pachylops chloropterus, *Fieb. Eur. Hem.* 285.

Germany. France.

Div. 13.

Dasycytus, *Fieb. Wien. Ent. Mon.* viii. 84, pl. 2, f. 11.

80. EURYMEROCORIS SORDIDUS.

Dasycytus sordidus, *Fieb. Wien. Ent. Mon.* viii. 85.

Malaga.

Div. 14.

Hypsitylus, *Fieb. Eur. Hem.* 286.

81. EURYMEROCORIS PRASINUS.

Hypsitylus prasinus, *Pict. Mey. Fieb. Eur. Hem.* 286.

South Spain.

Div. 15.

Platycranus, *Fieb. Verh. Zool. Bot. Ges. Wien.* xx. 252, pl. 6, f. 9.

82. EURYMEROCORIS ERBERI.

Platycranus Erberi, *Fieb. Verh. Zool. Bot. Ges. Wien.* xx. 252.

Dalmatia. Montenegro.

Div. 16.

Campotylus, *Fieb. Eur. Hem.* 70, 286.

83. EURYMEROCORIS YERSINI.

Capsus Yersini, *Muls. Ann. Soc. Linn.* 1856, Sp. 129—Camptotylus Yersini, *Fieb. Eur. Hem.* 287.

South France. South Spain.

84. EURYMEROCORIS MEYERI.

Camptotylus Meyeri, *Frey-Gessner, Mitthiel. Schweif. Ert-Gesellsch.* 1863, 119.

Sarepta.

Div. 17.

Loxops, *Fieb. Crit. Gen.* 54, pl. 6, f. 21; *Eur. Hem.* 70, 287.

85. EURYMEROCORIS COCCINEUS.

Capsus coccineus, *Westerh. Meyer, Caps.* Sp. 48, pl. 4, f. 5—Loxops coccineus, *Fieb. Eur. Hem.* 287.

Europe.

Div. 18.

Tichorhinus, *Fieb. Crit. Gen.* 55, pl. 6, f. 24—Litocoris, *Fieb. Eur. Hem.* 70, 287—Litosoma, p., *Dougl. and Scott, Hem.* 334.

86. EURYMEROCORIS ERICETORUM.

Phytocoris ericetorum, *Fall. Hem. Suec.* i. 105—Capsus ericetorum, *Sahlb. Geoo. Fen.* 104. *Kirschb. Caps.* 90. *Flor, Rhyn. Liv.* i. 587— Capsus limbatus? *Muls. Ann. Soc. Linn.* 1857, 165--Litocoris erice- torum, *Fieb. Eur. Hem.* 287—Litosoma ericetorum, *Dougl. and Scott, Hem.* 343.

Siberia.

a. Europe.   From Mr. Children's collection.

87. EURYMEROCORIS BICOLOR.

Litosoma bicolor, *Dougl. and Scott, Ent. M. Mag.* iv. 267, pl. 2, f. 3.
England.

88. EURYMEROCORIS VIRESCENS.

Litosoma virescens, *Dougl. and Scott, Hem.* 339.
England.

89. EURYMEROCORIS OCHROTRICHUS.

ochrotrichus, *Fieb. MSS. Dougl. and Scott, Hem.* 342.
England.

90. EURYMEROCORIS ANNULICORNIS.

Litocoris? annulicornis, *Synt. A. S. E. F. 4me Sér.* v. 126.
South France.

Div.?

91. EURYMEROCORIS QUADRIMACULATUS.

Phytocoris qnadrimaculatus, *Fall. Hem. Suec.* 119.
Sweden.   Siberia.

92. EURYMEROCORIS FLAVEOLUS.

flaveolus, *Stal, Stett. Ent. Zeit.* xix. 189.
Siberia.

93. EURYMEROCORIS OBSCURICEPS.
obscuriceps, *Stal, Stett. Ent. Zeit.* xix. 190.
Siberia.

Madeira.
94. EURYMEROCORIS WHITEI.
Phytocoris? Whitei, *Wltn. Ann. Nat. Hist.* 3rd *Ser.* i. 124.
a—h. Madeira. From Mr. Wollaston's collection.

Africa.
95. EURYMEROCORIS TABESCENS.
Capsus (Eurymerocoris) tabescens, *Stal, Ofv. Vet. Ak. Forh.* 1858, 317—
Eurymerocoris tabescens, *Stal, Hem. Afr.* iii. 22.
South Africa.

96. EURYMEROCORIS VIRIDIPUNCTATUS.
Capsus (Eurymerocoris) viridipunctatus, *Stal, Ofv. Vet. Ak. Forh.* 1858,
317—Eurymerocoris viridipunctatus, *Stal, Hem. Afr.* iii. 23.
South Africa.

Genus 8. MONALOCORIS.
Monalocoris, *Dahlb. Vet. Acad. Handl.* 1851. *Kirschb. Caps.* 31. *Fieb.
Crit. Gen.* 2; *Eur. Hem.* 61, 237. *Dougl. and Scott, Hem.* 278.

1. MONALOCORIS FILICIS.
Cimex Filicis, *Linn. Syst. Nat.* 718; *Faun. Suec.* 919—Acanthia Filicis,
Wolff, Wanz. Ins. f. 172. *Ic. Cim.* 46, pl. 5, f. 43—Phytocoris Filicis,
Fall. Hem. Suec. i. 108. *Hahn, Wanz. Ins.* ii. 86, pl. 56, f. 172—
Capsus Filicis, *Meyer, Dür. Caps.* 71. *Sahlb. Geoc. Fen.* 113.
Kirsch. Caps. 70—Bryoeoris Filicis, *Flor, Rhyn. Liv.* 539—Monalo-
coris Filicis, *Fieb. Eur. Hem.* 238. *Dougl. and Scott, Hem.* 279.
Europe.

2. MONALOCORIS BIPUNCTIPENNIS.
*Testaceus, ellipticus, glaber, sublus nigricante marginatus; caput latum;
oculi non prominuli; rostrum coxas posticas attingens; antennæ
gracillimæ, corpore paullo breviores; abdominis dorsum nigricans;
corii costa nigro biguttata; membrana diaphana.* Var. β.—*Prothorax et scutellum nigra.*

Testaceous, elliptical, smooth, shining, blackish along each side and
about the tip of the abdomen beneath. Head broad, short. Eyes not
prominent. Rostrum extending to the hind coxæ. Antennæ very slender,

a little shorter than the body.   Abdomen blackish above.   Legs short,
slender.   Corium with two black costal dots, one apical, the other subapical.   Membrane pellucid.   *Var. β.*—Prothorax and scutellum black.
Length of the body 1 line.

a—e. Ceylon.   Presented by Dr. Thwaites.

Genus 9. BRYOCORIS.

Bryocoris, *Fall. Hem. Suec.* 151.   *Fieb. Crit. Gen.* 3; *Eur. Hem.* 61, 238.
   Dougl. and Scott, Hem. 276.

1. BRYOCORIS PTERIDIS.

Capsus Pteridis, *Fall. Mon. Caps.* 105.   *Germ. Ahrms. Faun. Eur.* 10, pl.
   13.   *Meyer, Dür. Caps.* 114.   *Kirschb. Caps.* 76—Bryocoris Pteridis,
   Fall. Hem. Suec. i. 152.   *Zett. Ins. Lapp.* 266.   *Sahlb. Geoc. Fen.*
   124.   *Kol. Mel. Ent.* ii. 129.   *Fieb. Eur. Hem.* 238.   *Dougl. and
   Scott, Hem.* 277—Halticus Pteridis, *Burm. Handb. Ent.* ii. 278—
   Capsus pulcher, *Sahlb. Geoo. Fen.* 93.

a—c. England.   From Mr. Stephens' collection.
d. England.   Presented by F. Walker, Esq.
e. Europe.   From Mr. Children's collection.

Genus 10. FULVIUS.

Fulvius, *Stal, Stett. Ent. Zeit.* xxiii. 322.

1. FULVIUS ANTHOCOROIDES

anthocoroides, *Stal, Stett. Ent. Zeit.* xxiii. 322.
Mexico.

Genus 11. ANAPUS.

Anapus, *Stal, Stett. Ent. Zeit.* xix. 188.

1. ANAPUS KIRSCHBAUMI.

Kirschbaumi, *Stal, Stett. Ent. Zeit.* xix. 189.
Siberia.

Genus 12. MYRMECORIS.

Myrmecoris, *Gorsk. Ann. Ent. Imp. Ross.*   *Fieb. Crit. Gen.* 4; *Eur.
   Hem.* 61, 238.

1. MYRMECORIS GRACILIS.

Globiceps gracilis, *Sahlb. Geoc. Fen.* 123.   *K. V. Acad. Forh.* 1852—
   Myrmecoris agilis, *Gorsk. Ann.* 167, pl. 2, f. 1.   (M. lituanica)
   Kirschb. Rhyn. 40—Myrmecoris gracilis, *Fieb. Eur. Hem.* 239.
Europe.

Genus 13. PSILORHAMPHUS.

Psilorhamphus, *Stal, Ofv. K. V. Ak. Forh.* 1870, 669.

1. PSILORHAMPHUS CONSPERSUS.

conspersus, *Stal, Ofv. K. V. Ak. Forh.* 1870, 669.
Philippine Isles.

2. PSILORHAMPHUS CONSPUTUS.

conspntus, *Stal, Ofv. K. V. Ak. Forh.* 1870, 670.
Philippine Isles.`   •

3. PSILORHAMPHUS ALBOMACULATUS.

albomaculatus, *Stal, Ofv. K. V. Ak. Forh.* 1870, 670.
Philippine Isles.

Genus 14. DISPHINCTUS.

Disphinctus, *Stal, Ofv. K. V. Ak. Forh.* 1870, 668.

1. DISPHINCTUS FALLENII.

Fallenii, *Stal, Ofv. K. V. Ak. Forh.* 1870, 668, pl. 7, f. 6.
Philippine Isles.

2. DISPHINCTUS SAHLBERGI.

Sahlbergi, *Stal, Ofv. K. V. Ak. Forh.* 1870, 668.
Philippine Isles.

3. DISPHINCTUS REUTERI.

Reuteri, *Stal, Ofv. K. V. Ak. Forh.* 1870, 668.
Philippine Isles.

4. DISPHINCTUS HAGLUNDII.

Haglundii, *Stal, Ofv. K. V. Ak. Forh.* 1870, 668.
Philippine Isles.

Genus 15. MONALONION.

Monalonion, *H.-Sch. Wanz. Ius.* ix. 168.

1. MONALONION PARVIVENTRE.

parviventre, *H.-Sch. Wanz. Ins.* ix. 168, pl. 312, f. 958.
Brazil.

2. MONALONION SCHÆFFERI.

Schæfferi, *Stal, Rio Jan. Hem.* 56.
Rio Janeiro.

3. Monalonion annulipes.

annulipes, *Sgnt. A. S. E. F. 3me Sér.* vi. 500.

Mexico.

4. Monalonion braconoides.

Mas. Nigrum, glabrum, fere lineare ; oculi valde prominuli ; antennæ piceæ, gracillimæ, corpore multo longiores, articulo 1o crasso brevi basi rufo, 2o pubescente ; prothorax, abdomen basi et pedes quatuor anteriores rufa ; femora postica nigro fasciata ; alæ nigricantes, semihyalinæ, nigro venosæ.

Male. Black, smooth, shining, slender, nearly linear. Head short. Eyes very prominent. Rostrum tawny, extending to the middle coxæ. Antennæ piceous, very slender, much longer than the body; first joint thick, shorter than the head, red towards the base; second pubescent, full six times as long as the first ; third much shorter than the second. Prothorax red, convex, contracted and forming a neck in front. Scutellum red. Abdomen red towards the base. Legs long. Four anterior legs red, slender. Hind legs black, stouter; hind femora with a yellow band beyond the middle. Wings blackish, semihyaline, extending much beyond the tip of the abdomen ; veins black; areolet of the membrane very large. Length of the body 3½ lines.

a. Amazon Region. From Mr. Bates' collection.

5. Monalonion ichneumonoides.

Nigrum ; antennæ pubescentes, corpore multo longiores, articulo 1o crasso brevissimo basi rufo, 2o rufescente ; prothoracis lobus posticus, scutellum, abdomen basi pedesque quatuor anteriores rufa ; femora postica flavo unifasciata ; alæ nigricantes, abdomen longe superantes.

Black, slender, smooth, shining. Head transverse, short, red about the mouth. Eyes prominent. Antennæ slender, pubescent, much longer than the body ; first joint thick, very short, red at the base; second reddish, much longer than the third. Prothorax red above, excepting the neck. Scutellum red. Abdomen red towards the base. Legs slender, rather long. Four anterior legs red ; hind legs black, their femora with a yellow band at somewhat beyond the middle. Wings blackish, extending much beyond the tip of the abdomen. Length of the body 3 lines.

a. Amazon Region. From Mr. Bates' collection.

Asia and Eastern Isles.

6. Monalonion humerale.

Rufum, gracile, subtilissime punctatum ; caput breve ; oculi prominuli ; rostrum coxas intermedias attingens ; antennæ nigræ, gracillimæ, articulo 1o piceo sat valido ; prothorax antice arctatus et bisulcatus, postice nigro bimaculatus ; pedes lutei ; alæ anticæ fuscescente cinereæ, fusco venosæ.

Red, slender, shining, very finely punctured. Head short-triangular. Eyes black, prominent. Rostrum extending to the middle coxæ. Antennæ black, very slender; first joint piceous, rather stout, a little shorter than the head; second more than thrice the length of the first. Prothorax contracted in front, with two transverse furrows, and with a large black spot on each side hindward. Legs luteous, slender. Corium and membrane hyaline, brownish cinereous; veins brown. Length of the body 4 lines.

a. Malacca. Presented by W. W. Saunders, Esq.

7. MONALONION POLITUM.

Fœm. *Rufum, gracile, fere lineare; caput breve; oculi valde promi-nuli; rostrum coxas anticas attingens; antennæ nigræ, articulo 1o piceo valdo capiti æquilongo; prothorax antice arctatus et transverse bisulcatus, angulis posticis acutis; corium apice fuscum; membrana fusca.*

Female. Red, slender, smooth, shining, nearly linear. Head short-triangular. Eyes piceous, very prominent. Rostrum extending to the fore coxæ. Antennæ black, slender; first joint piceous, stout, as long as the head; second more than four times the length of the first. Prothorax contracted in front, where there are two well-defined transverse furrows; hind angles acute. Legs slender. Corium brown at the tip. Membrane and hind wings brown. Length of the body 4 lines.

a. Sarawak. Presented by W. W. Saunders, Esq.

9. MONALONION DIVISUM.

Mas. *Nigrum, gracile, sublineare; caput rufum, nigro unifasciatum; oculi prominuli; rostrum coxas posticas paullo superans; antennæ corpore paullo breviores; prothorax antice bisulcatus, postice striatus; abdomen basi rufum; pedes rufi; corium rufo late unifasciatum.*

Male. Black, slender, nearly linear. Head red, triangular, smooth, shining, with a black band on the hind part of the vertex. Eyes black, prominent. Rostrum red, extending a little beyond the hind coxæ. Antennæ shorter than the body; first joint as long as the head; second about thrice the length of the first; third as long as the first. Prothorax red, with two transverse furrows in front and with a broad hinder band, which is transversely striated. Abdomen red towards the base. Legs red, slender. Corium with a broad red band at one-third of the length. Membrane blackish. Length of the body 4½ lines.

a. Ternate. From Mr. Wallace's collection.

Genus 16. EUCEROCORIS.

Eucerocoris, *Westw. Trans. Ent. Soc. Lond.* ii. 21.

1. EUCEROCORIS NIGRICEPS.

nigriceps, *Westw. Trans. Ent. Soc. Lond.* ii. 22, pl. 2, f. 7.

2. EUCEROCORIS WESTWOODII.

Westwoodii, *White, Trans. Ent. Soc. Lond.* iii. 94.
Sierra Leone.

3. EUCEROCORIS BRACONOIDES.

Fœm. *Niger, gracilis, glaber; caput rufum, nigricante uniguttatum, antice conicum; rostrum coxas anticas attingens; antennæ hirsutæ, corpore longiores; prothoracis lobus anticus rufus, bicallosus; lobi postici latera rufa; scutellum rufo univittatum; abdomen album, fasciis quatuor apiceque nigris; pedes longi, hirsuti, femoribus supra flavis; alæ nigricantes; corii costa rufo strigata.*

Female. Black, slender, smooth, shining, nearly linear. Head and fore lobe of the prothorax red. Head conical in front; a blackish dot between the eyes. Eyes black, prominent. Rostrum extending to the fore coxæ. Antennæ black, slender, hirsute, longer than the body; second joint much longer than the first. A callus on each side behind the neck, which is well defined and has a transverse furrow. Hind lobe of the prothorax red on each side. Scutellum with a red stripe. Abdomen white, with a black band near the base, with three black bands on the hinder half and with a black tip. Legs long, slender, hirsute; femora yellow above. Wings blackish; corium with a red costal streak near the base. Length of the body 4 lines.

a. W. Australia. From M. Du Boulay's collection.

4. EUCEROCORIS BASIFER.

Fœm. *Niger, gracilis, setulosus, sublinearis; caput subtus, prothorax, scutellum, pectus, coxæ et alæ anticæ basi coccinea; vertex rufo bimaculatus; oculi prominuli; rostrum rufum, coxas anticas attingens; antennæ corpore longiores, basi rufæ, articulo 1o apicem versus subincrassato; prothorax antice arctatus, sulco transverso bene determinato; corium basi albo unistrigatum; membrana nigricans.*

Female. Black, slender, setulose, shining, nearly linear. Head beneath, prothorax, scutellum, pectus, coxæ and fore wings at the base bright red. Head transverse, with a red spot on the vertex. Eyes prominent. Rostrum red, extending to the fore coxæ. Antennæ and legs setulose, slender. Antennæ longer than the body; first joint longer than the prothorax, red at the base, slightly incrassated towards the tip; second much longer than the first. Prothorax straightened in front, with a strongly-marked transverse furrow near the fore border. Corium with a white streak near the base. Membrane and hind wings blackish. Length of the body 5 lines.

a, b. Australia. From Mr. Argent's collection.

Genus 17. PACHYPELTIS.

Pachypeltis, *Sgnt. A. S. E. F. 3me Sér.* vi. 501.

1. PACHYPELTIS CHINENSIS.

Chinensis, *Sgnt. A. S. E. F. 3me Sér.* vi. 501.
China.

Genus 18. SPHINCTOTHORAX.

Sphinctothorax, *Stal, Ofv. Vet. Ak. Forh.* 1853, 260; *Hem. Afr.* iii. 17.

1. SPHINCTOTHORAX LEUCOPHÆUS.

Cyllocoris leucophæus, *Germ. Silb. Rev. Ent.* v. 135—Sphinctothorax leucophæus, *Stal, Hem. Afr.* iii. 18.
Caffraria.

Genus 19. HELOPELTIS.

Helopeltis, *Sgnt. A. S. E. F.* 3me *Sér.* vi. 502—Aspicelus, *Costa, Ann. Mus. Zool. Nap.* ii. 146.

1. HELOPELTIS ANTONII.

Antonii, *Sgnt. A. S. E. F.* 3me *Sér.* vi. 502, pl. 12, f. 2.
Ceylon.

2. HELOPELTIS PODAGRICUS.

Aspicelus podagricus, *Costa, Ann. Mus. Zool. Nap.* ii. 147, pl. 2, f. 6.

3. HELOPELTIS NIGER.

Mas. *Niger; rostrum coxas intermedias attingens; antennæ gracillimæ, corpore plus duplo longiores; scutelli cornu capitatum, erectum; alæ nigricantes.*

Male. Black, slender, shining, nearly linear. Head small, short. Eyes very prominent. Rostrum extending to the middle coxæ. Antennæ very slender, more than twice the length of the body; first joint rather shorter than the body; second much longer than the first. Prothorax contracted in front, where there is a transverse furrow. Scutellum with a long filiform capitate appendage. Legs long, slender. Wings blackish, slightly hyaline. Length of the body 2½ lines.

a. Wagiou. Presented by W. W. Saunders, Esq.

4. HELOPELTIS BRACONIFORMIS.

Niger; rostrum coxas posticas attingens, articulo 1o rufescente; antennæ corpore multo longiores, articulo 1o subclavato; prothorax et scutellum rufa, hujus spina nigra capitata erecta; abdamen basi albidum; pedes rufescentes, femoribus subnodulosis nigro trifasciatis; alæ anticæ basi albidæ.

Black, slender, smooth, shining. Head transverse, very short. Eyes prominent. Rostrum extending to the hind coxæ; first joint reddish. Antennæ very slender, much longer than the body; first joint subclavate; second much longer than the first; third a little shorter than the first. Prothorax and scutellum red, the latter with a slender black erect capitate horn. Abdomen whitish at the base. Legs reddish, long, slender; femora slightly nodulose, black towards the tips and with two black bands. Wings

blackish, extending much beyond the abdomen. Fore wings membranaceous, whitish at the base. Length of the body 3 lines.

a. New Guinea. Presented by W. W. Saunders, Esq.

Genus 20. HERDONIUS.

Herdonius, *Stal, Rio Jan. Hem.* 55.

1. HERDONIUS ARMATUS.

armatus, *Stal, Rio Jan. Hem.* 56.
Rio Janeiro.

Genus 21. VALDASUS.

Valdasus, *Stal, Rio Jan. Hem.* 56.

1. VALDASUS FAMULARIS.

famularis, *Stal, Stett. Ent. Zeit.* xxiii. 321.
Mexico.

2. VALDASUS SCHONHERRI.

Schonherri, *Stal, Rio Jan. Hem.* 56.
Rio Janeiro.

Genus 22. ECCRITOTARSUS.

Eccritotarsus, *Stal, Rio Jan. Hem.* 57.

Mexico.

A. Tarsi brown. - - - - - - generosus.
B. Tarsi whitish. - - - - - eucosmus.
C. Tarsi pale yellow.
{a. First joint of the rostrum short. - - - pallidirostris.
b. First joint of the rostrum long. - - - mundulus.

1. ECCRITOTARSUS GENEROSUS.

generosus, *Stal, Stett. Ent. Zeit.* xxiii. 323.
Mexico.

2. ECCRITOTARSUS EUCOSMUS.

eucosmus, *Stal, Stett. Ent. Zeit.* xxiii. 323.
Mexico.

3. ECCRITOTARSUS PALLIDIROSTRIS.

pallidirostris, *Stal, Stett. Ent. Zeit.* xxiii. 323.
Mexico.

4. ECCRITOTARSUS MUNDULUS.

mundulus, *Stal, Stett. Ent. Zeit.* xxiii. 323.
Mexico.

South America.

A. Head more or less sloping before the eyes.
a. Body black or blackish.
    * Corium pale at the base.   -   -   -   pallidipes.
    ** Corium wholly blackish.   -   -   -   dimidiatus.
    *** Corium pale for half the length from the base. -   nigrocruciatus.
    **** Corium red. -   -   -   -   -   Fairmairei.
    ***** Corium black, with a whitish apical spot.   -   venustus.
    ****** Corium black, with two whitish spots.   -   leucopus.
    ******* Corium whitish, with a brown band. -   -   discipennis.
    ******** Corium yellowish, with a large brown spot.   -   longulus.
b. Body pale.
    * Corium black, pale at the base.   -   -   semiluteus.
    ** Corium wholly pale. -   -   -   -   lutescens.
    *** Corium reddish, with a large blackish spot.   -   discifer.
    **** Corium blackish behind the middle. -   -   cruxnigra.
    ***** Corium with the interior half blackish.   - nigroplagiatus.
    ***** Corium with a blackish band. -   -   -   hyalinus.
B. Head elongated and conical in front. -   -   niger.

5. ECCRITOTARSUS SEMILUTEUS.

semiluteus, *Stal, Rio Jan. Hem.* 57.
Rio Janeiro.

6. ECCRITOTARSUS PALLIDIPES.

pallidipes, *Stal, Rio Jan. Hem.* 57.
Rio Janeiro.

7. ECCRITOTARSUS DIMIDIATUS.

dimidiatus, *Stal, Rio Jan. Hem.* 57.
Rio Janeiro.

8. ECCRITOTARSUS LUTESCENS.

lutescens, *Stal, Rio Jan. Hem.* 57.
Rio Janeiro.

9. ECCRITOTARSUS NIGROPLAGIATUS.

nigroplagiatus, *Stal, Rio Jan. Hem.* 57.
Rio Janeiro.

10. ECCRITOTARSUS DISCIFER.

discifer, *Stal, Rio Jan. Hem.* 57.
Rio Janeiro.

11. ECCRITOTARSUS NIGROCRUCIATUS.

nigrocruciatus, *Stal, Rio Jan. Hem.* 57.
Rio Janeiro.

12. ECCRITOTARSUS CRUXNIGRA.

cruxnigra, *Stal, Rio Jan. Hem.* 58.
Rio Janeiro.

13. ECCRITOTARSUS FAIRMAIREI.

Fairmairei, *Stal, Rio Jan. Hem.* 58.
Rio Janeiro.

14. ECCRITOTARSUS VENUSTUS.

venustus, *Stal, Rio Jan. Hem.* 58.
Rio Janeiro.

15. ECCRITOTARSUS LEUCOPUS.

leucopus, *Stal, Rio Jan. Hem.* 58.
Rio Janeiro.

16. ECCRITOTARSUS DISCIPENNIS.

discipennis, *Stal, Rio Jan. Hem.* 58.
Rio Janeiro.

17. ECCRITOTARSUS HYALINUS.

hyalinus, *Stal, Rio Jan. Hem.* 58.
Rio Janeiro.

18. ECCRITOTARSUS LONGULUS.

longulus, *Stal, Rio Jan. Hem.* 58.
Rio Janeiro.

19. ECCRITOTARSUS NIGER.

niger, *Stal, Rio Jan. Hem.* 58.
Rio Janeiro.

20. ECCRITOTARSUS FULVICOLLIS.

Capsus fulvicollis, *Fabr. Syst. Rhyn.* 244—Eccritotarsus fulvicollis, *Sta*
  Hem. Fabr. i. 85.
South America.

Genus 23. SINERVUS.

Sinervus, *Stal, Rio Jan. Hem.* 56.

1. SINERVUS BARENSPRUNGI.

Bärensprungi, *Stal, Rio Jan. Hem.* 56.
Rio Janeiro.

Genus 24. AMBRACIUS.

Ambracius, *Stal, Rio Jan. Hem.* 59.

1. AMBRACIUS DUFOURI.

Dufouri, *Stal, Rio Jan. Hem.* 59.
Rio Janeiro.

2. Ambracius phaleratus.

phaleratus, *Stal, Rio Jan. Hem.* 59.
Rio Janeiro.

Genus 25. HENICOCNEMIS.

Henicocnemis, *Stal, Rio Jan. Hem.* 53.

1. Henicocnemis albitarsis.

albitarsis, *Stal, Stett. Ent. Zeit.* xiii. 320.
Mexico.

2. Henicocnemis patellata.

patellata, *Stal, Rio Jan. Hem.* 53.
Rio Janeiro.

The following genus is stated to belong to this family.

Genus CALLIPREPES.

Calliprepes, *White, Trans. Ent. Soc. Lond.* iii. 93.

Calliprepes Grayii.

Grayii, *White, Trans. Ent. Soc. Lond.* iii. 94.
Nepaul.

DUCTIROSTRA.

A. Fore legs raptorial; femora very thick. - - 1. Spissipedes.
B. Fore legs not raptorial.
a. Body generally with membranaceous appendages.
   Fore wings usually longer and broader than the
   abdomen, with reticulated veins. - - 2. Membranacea.
b. Body flat. Fore wings generally shorter and
   narrower than the abdomen, with irregular
   veins. - - - - - - 3. Corticola.
c. Groove for the rostrum extremely slight. Fourth
   joint of the antennæ setiform. - - - 4. Lecticola.

The Lecticola are more allied to the Nudirostra than to the Ductirostra, and have much affinity to the Anthocoridæ and Microphysidæ. The Ripicola are associated in this arrangement with the Remipeda.

1. SPISSIPEDES.

A. Head elongated and bifid between the antennæ.
   Scutellum very short. - - - 1. Phymatidæ.
B. Head not elongated. Scutellum long. - 2. Macrocephalidæ.

Fam. 1. PHYMATIDÆ.

Phymatidæ, *Serv. Hist. Hem.* 288. *Fieb. Eur. Hem.* 24.

Fieber includes the Macrocephalidæ with the Phymatidæ, and notices the four following genera :—

A. Head thick, quadrangular on the sides, elongated in front.   -   -   - PHYMATA.

B. Head compressed on the sides, almost cylindrical, truncated in front.

a. Fourth joint of the antennæ slender, as long as the three first together.   -   - MECODACTYLUS, *Fieber.*

b. Fourth joint of the antennæ stout, a little longer than the second and third together.

* Scutellum not extending to the tip of the abdomen.   -   -   -   - CARCINOCHELIS, *Fieber.*

** Scutellum extending to the tip of the abdomen.   -   -   -   -   - MACROCEPHALUS.

Genus 1. PHYMATA.

Phymata, *Latr. Gen.* iii. 138. *Serv. Hist Hem.* 288. *Fieb. Eur. Hem.* 33—
Syrtis, *Fabr. Syst. Rhyn.* 123. *Burm. Handb. Ent.* ii. 251.

Europe.

1. PHYMATA CRASSIPES.

—— ——, *Geoffr. Ins.* i. 447. *Schæff. Icon.* pl. 57, f. 12—Cimex crassipes, *Linn. Syst. Nat.* i. 2. *Rossi, Faun. Etr.* 4, 1286—Syrtis crassipes, *Fabr. Syst. Rhyn.* 121. *Schelb. Wanz. Ins.* 4, pl. 6, f. 3. *Hahn, Wanz. Ins.* iii. 58. *Burm. Handb. Ent.* ii. 251. *Ramb. Faun. And.* ii. 167—Cimex crassipes, *Panz. Faun. Germ.* 23, 24—Acanthia crassipes, *Wolff. Icon. Cim.* 88, pl. 9, f. 82. *Coqueb. Ill. Icon.* 31, pl. 21, f. 6—Phymata crassipes, *Latr. Gen.* iii. 138. *St. Farg. et Serv. Enc. Méth.* x. 119. *L. Duf. Rech. Hém.* 53. *Brullé, Hist. Nat. Ins.* ix. 347. *Blanch. Hist. Nat. Ins.* iii. 114. *Westw. Trans. Ent. Soc. Lond.* iii. pl. 2, f. 2. *Serv. Hist. Hem.* 290. *Fieb. Eur. Hem.* 110.

Europe.

2. PHYMATA MONSTROSA.

Acanthia monstrosa, *Fabr.*—Syrtis monstrosa, *Fabr. Syst. Rhyn.* 122. *Burm. Handb. Ent.* ii. 251. *Hahn, Wanz. Ins.* iii. 57, pl. 90, f. 273— Phymata monstrosa, *Fieb. Eur. Hem.* 110.

a, b. Europe. Presented by W. W. Saunders, Esq.

North America.

3. PHYMATA EROSA.

Cimex erosa, *Linn. Syst. Nat.* 2, 718. *Tign. Hist. Nat. Ins.* iv. 264, pl. 6, f. 3—Cimex scorpio, *Deg. Ins.* iii. 350, pl. 35, f. 13—Syrtis erosa,

Fabr. Syst. Rhyn. 121. *Enc. Méth.* pl. 374, f. 6. *H.-Sch. Wanz. Ins.* vii. 15, pl. 222, f. 694—Acanthia erosa, *Fabr. Ent. Syst.* iv. 74. *Wolff, Icon. Cim.* 89, pl. 9, f. 83—Phymata erosa, *St. Farg. et Serv. Enc. Méth.* x. 119. *De Lap. Hem.* 14, pl. 51, f. 4. *Westw. Trans. Ent. Soc. Lond.* iii. pl. 2, f. 3.

a, b. United States. Presented by E. Doubleday, Esq.
c. United States. Presented by W. W. Saunders, Esq.
d. United States. Presented by F. Walker, Esq.

4. PHYMATA CARNEIPES.

carneipes, *Mayr, Verh. Zool. Bot. Ges. Wien.* xv. 442. *Reise Novara, Zool.* ii. *Hem.* 168, pl. 5, f. 49.

Georgia. Brazil.

Mexico.

5. PHYMATA ANNULIPES.

annulipes, *Stal, Stett. Ent. Zeit.* xxiii. 439.
Mexico.

West Indies.

6. PHYMATA MARGINATA.

Syrtis marginata, *Fabr. Syst. Rhyn.* 122—Phymata marginata, *Stal, Hem. Fabr.* i. 93.

West Indies.

7. PHYMATA EMARGINATA.

emarginata, *Guér. Hist. Fis. Cuba,* vii. 169.
Cuba.

8. PHYMATA ACUTANGULA.

acutangula, *Guér. Hist. Fis. Cuba,* vii. 169.
Cuba.

9. PHYMATA CARINATA.

Syrtis carinata, *Fabr. Syst. Rhyn.* 122—Phymata carinata, *Spin. Faun. Chil.* 206, pl. 2, f. 12. *Stal, Hem. Fabr.* i. 93.

West Indies. South America. Chili.

South America.

10. PHYMATA SPINOSISSIMA.

spinosissima, *Mayr, Verh. Zool. Bot. Gesell. Wien.* xv. 442.
Brazil.

11. PHYMATA SIMULANS.

simulans, *Stal, Rio Jan. Hem.* 59.
Rio Janeiro.

12. PHYMATA ACUTA.

acuta, *Stal, Rio Jan. Hem.* 60.
Rio Janeiro.

13. PHYMATA LONGICEPS.

longiceps, *Stal, Rio Jan. Hem.* 59.
Rio Janeiro.

14. PHYMATA FASCIATA.

fasciata, *Stal, Rio Jan. Hem.* 59.
Rio Janeiro.

15. PHYMATA SWEDERI.

Swederi, *Stal, Rio Jan. Hem.* 60.
Rio Janeiro.

16. PHYMATA NERVOSOPUNCTATA.

nervosopunctata, *Sgnt. A. S. E. F. 4me Sér.* iii. 574, pl. 13, f. 25.
Chili.

17. PHYMATA ELONGATA.

elongata, *Sgnt. A. S. E. F. 4me Sér.* iii. pl. 13, f. 26.
Chili.

New Zealand.

18. PHYMATA FEREDAYI.

Feredayi, *Scott, Stett. Ent. Zeit.* xxiii. 102.
New Zealand.

19. PHYMATA CONSPICUA.

conspicua, *Scott, Stett. Ent. Zeit.* xxiii. 102.
New Zealand.

Country unknown.

20. PHYMATA INTEGRA.

integra, *Westw. Trans. Ent. Soc. Lond.* iii. 22, pl. 2, f. 1.

Fam. 2. MACROCEPHALIDÆ.

Macrocephalidæ, *Serv. Hist. Hem.* 296.

A. Scutellum extending to the middle of the abdomen.
  a. Scutellum acute at the tip. -   -   - 1. OXYTHYREUS.
  b. Scutellum rounded at the tip. -   -   - 2. AMBLYTHYREUS.
B. Scutellum extending to the tip of the abdomen. - 3. MACROCEPHALUS.

Genus 1. OXYTHYREUS.

Oxythyreus, *Westw. Trans. Ent. Soc. Lond.* iii. 27. *Serv. Hist. Hem.* 291.

1. OXYTHYREUS CYLINDRICORNIS.

Macrocephalus (Hemithyreus) cylindricornis, *Westw. Trans. Ent. Soc. Lond.* iii. 28, pl. 11, f. 7. *Serv. Hist. Hem.* 291.

Country unknown.

Genus 2. AMBLYTHYREUS.

Amblythyreus, *Westw. Trans. Ent. Soc. Lond.* iii. 28.

1. AMBLYTHYREUS QUADRATUS.

M. (Amblythyreus) quadratus, *Westw. Trans. Ent. Soc. Lond.* iii. 31.

Hindostan.

2. AMBLYTHYREUS RHOMBIVENTRIS.

M. (Amblythyreus) rhombiventris, *Westw. Trans. Ent. Soc. Lond.* iii. 30, pl. 2, f. 8.

3. AMBLYTHYREUS ANGUSTUS.

angustus, *Westw. Trans. Ent. Soc. Lond.* iii. 31.

Genus 3. MACROCEPHALUS.

Macrocephalus, *Swed. Nov. Act. Holm.* 1787, iii. *Latr. Gen.* iii. 137. *Serv. Hist. Hem.* 292. *Burm. Handb. Ent.* ii. 552.

North America.

1. MACROCEPHALUS CIMICOIDES.

Macrocephalus cimicoides, *Swed. Nov. Act. Holm.* 1787, iii. pl. 8, f. 1 *Latr. Gen.* iii. 137, 376. *Westw. Trans. Ent. Soc.* iii. 23, pl. 2, f. 5—Syrtis manicata, *Fabr. Syst. Rhyn.* 123—Macrocephalus manicatus, *Burm. Handb. Ent.* ii. 252.

Carolina. Georgia.

2. MACROCEPHALUS PREHENSILIS.

Syrtis prehensilis, *Fabr. Syst. Rhyn.* 123—Macrocephalus prehensilis, *Westw. Trans. Ent. Soc.* iii. 26. *Serv. Hist. Hem.* 293—Macrocephalus pallidus, *Westw. Trans. Ent. Soc. Lond.* iii. 27. *H.-Sch. Wanz. Ins.* viii. 108, pl. 285, f. 879.

North America. Brazil.

3. MACROCEPHALUS MACILENTUS.
macilentus, *Westw. Trans. Ent. Soc. Lond.* iii. 27, pl. 2, f. 6.
North America. Colombia.

Mexico.
4. MACROCEPHALUS INCISUS.
incisus, *Stal, Stett. Ent. Zeit.* xxiii. 440.
Mexico.

5. MACROCEPHALUS CLIENS.
cliens, *Stal, Stett. Ent. Zeit.* xxiii. 440.
Mexico.

6. MACROCEPHALUS LEPIDUS.
lepidus, *Stal, Stett. Ent. Zeit.* xxiii. 440.
Mexico.

7. MACROCEPHALUS FALLENI.
Falleni, *Stal, Stett. Ent. Zeit.* xxiii. 441.
Mexico.

West Indies.
8. MACROCEPHALUS PULCHELLUS.
pulchellus, *Klug, Westw. Trans. Ent. Soc. Lond.* iii. 25.
Cuba.

9. MACROCEPHALUS RUGOSIPES.
rugosipes, *Guér. Hist. Ins. Cuba*, vii. 169.
Cuba.

10. MACROCEPHALUS WESTWOODII.
Westwoodii, *Guér. Hist. Ins. Cuba*, vii. 169.
Cuba.

South America.
11. MACROCEPHALUS LEUCOGRAPHUS.
leucographus, *Klug, Westw. Trans. Ent. Soc.* iii. 25.
Port au Prince.

12. MACROCEPHALUS NOTATUS.
notatus, *Westw. Trans. Ent. Soc. Lond.* iii. 24.
Colombia.

13. MACROCEPHALUS AFFINIS.
affinis, *Guér. Icon. R. An. Ins.* pl. 56, f. 10.   *Westw. Trans. Ent. Soc.
Lond.* iii. 26.
Brazil.

14. MACROCEPHALUS CRASSIMANUS.

Syrtis crassimana, *Fabr. Syst. Rhyn.* 123—Macrocephalus crassimanus,
Westw. Trans. Ent. Soc. Lond. iii. 26. *Serv. Hist. Hem.* 292, pl. 6, f. 2.
Stal, Hem. Fabr. i. 94.

South America.

15. MACROCEPHALUS TUBEROSUS.

tuberosus, *Klug, Trans. Ent. Soc. Lond.* iii. 24.

Cassapava.

16. MACROCEPHALUS OBSCURUS.

obscurus, *Westw. Trans. Ent. Soc. Lond.* iii. 24.

South America.

17. MACROCEPHALUS FORTIFICATUS.

Syrtis fortificata, *Mus. Berol. H.-Sch. Wanz. Ins.* vii. 15, pl. 222, f. 695.

Brazil.

MEMBRANACEA.

A. Scutellum covered by the prothorax. -   -   - 1. TINGIDIDÆ.
B. Scutellum not covered by the prothorax.   -   - 2. PIESMIDÆ.

Fam. 1. TINGIDIDÆ.

Tingididæ, *Serv. Hist. Hem.* 295.   *Fieb. Eur. Hem.* 24 — Tingidida
   (Agrammidæ et Tingididæ), *Dougl. and Scott, Hem.* 242.

A. Winged.
a. Antennæ not very long.
  * Scutellum *covered.*
  † Fore wings with no clavus.
  ‡ Third and fourth joints of the antennæ much
     incrassated.   -   -   -   -   1. DICTYONOTA.
  ‡‡ Third joint of the antennæ cylindrical.
  § Prothorax and fore wings with a vesiculose
     swelling. -   -   -   -   -   2. TINGIS.
  §§ No vesiculose swelling on the fore wings.
  ✕ Prothorax with three keels.
  o Head unarmed.   Prothorax forming hindward
     a much elongated triangle.   -   -   3. MONANTHIA.
  oo Head horned.   Prothorax almost rounded hind-
     ward.   -   -   -   -   -   4. CANTACADER.
  ✕✕ Prothorax with one keel.   -   -   -   5. AGRAMMA.
  †† Fore wings with a clavus.   -   -   - 6. TAPHROSTETHUS.
  ** Scutellum *uncovered.*
  † Spines of head long, acute.

‡ Abdomen not dilated.   -   -   -   7. Phatnoma.
‡‡ Abdomen much dilated.   -   -   -   8. Phyllotingis.
†† Spines of head short, obtuse. -   -   -   9. Teleia,
b. Antennæ very long. -   -   -   -   10. Tigava.
B. Wingless. -   -   -   -   - 11. Coleopterodes.

The following characters are some of those by which Fieber arranges
the European Tingididæ:—

A. Head elongated. -   -   -   -   -   Cantacader.
B. Head short, obtuse.
a. Sides of the prothorax not foliaceous.   -   -   Agramma.
b. Sides of the prothorax foliaceous.
  * Third joint of the antennæ clavate; fourth not in
      the axis of the third. -   -   -   -   Lacometopus.
** Third joint of the antennæ cylindric or filiform;
      fourth in the axis of the third.
  † Fore wings with forked veins.
  ‡ Head above with three or four areas.   -   -   Monanthia.
  ‡‡ Head above with two areas.
  § First and second joints of the antennæ thicker than
    ‖ the third.   -   -   -   -   -   Dictyonota.
  §§ First and second joints of the antennæ scarcely
      thicker than the third. -   -   -   -   Deryphysia.
  †† Fore wings with entire veins.
  ‡ First joint of the antennæ long, cylindric, from thrice
      to six times the length of the second.   -   -   Tingis.
  ‡‡ First joint of the antennæ short, thick, not more than
      twice the length of the second.
  § Neck with a rhomboidal vesicle.   -   -   -   Orthostira.
  §§ Neck with no vesicle.   -   -   -   -   Campylostira.

In this list Deryphysia is included with Dictyonota; Lacometopus,
Orthostira, Campylostira and Diconocoris with Monanthia; Galeatus with
Tingis.

Genus 1. DICTYONOTA.

Eurycera, *De Lap. Hém.* 49. *Burm. Handb. Ent.* ii. 258. *Brullé, Hist.
Nat. Ins.* ix. 341. *Blanch. Hist. Nat. Ins.* iii. 113. *Serv. Hist. Hem.*
295—Dictyonota, *Curt. Brit. Ent.* iv. 154. *Fieb. Wien. Ent. Mon.*
vii. f. 42; *Eur. Hem.* 126. *Dougl. and Scott, Hem.* 255.

Div. 1. Eurycera.

1. Dictyonota nigricornis.

Eurycera nigricornis, *De Lap. Hém.* 49. *Burm. Handb. Ent.* ii. 258.
Serv. Hist. Hem. 296, pl. 6, f. 3—Eurycera clavicornis, *H.-Sch. Wanz.
Ins.* iv. 65, pl. 129, f. 400. *Brullé, Hist. Nat. Ins.* ix. 341. *Blanch.
Hist. Nat. Ins.* iii. 113.

Europe.

a. Switzerland. Presented by F. Walker, Esq.

Div. 2. Dictyonota, *Curt. Fieb. Eur. Hem.* 126.

2. DICTYONOTA LUGUBRIS.

lugubris, *Fieb. Eur. Hem.* 126—fuliginosa? *Costa, Cent.* 1852, pl. 6, f. 5.
Servia.

3. DICTYONOTA ERYTHROPHTHALMA.

Tingis erythrophthalma, *Germ. Ahr. Faun. Eur.* 3, 25—Dictyonota erythrophthalma, *H.-Sch. Wanz. Ins.* iv. 74. *Fieb. Ent. Mon.* 94,
pl. 8, f. 1—3; *Eur. Hem.* 127.

a—d. England.

4. DICTYONOTA CRASSICORNIS.

Tingis crassicornis, *Fall. Mon. Cim.* 38; *Hem. Suec.* 147—Tingis pilicornis,
H.-Sch. Fanz. Faun. Germ. 118, 17—Dictyonota crassicornis, *Curt.
Brit. Ent.* iv. pl. 154. *Fieb. Ent. Mon.* 92, pl. 7, f. 42—47; *Eur. Hem.*
127. *Sahlb. Geoc. Fen.* 34. *H.-Sch. Wanz. Ins.* ix. 157. *Flor,
Rhyn. Liv.* i. 358. *Dougl. and Scott, Hem.* 255—Piesma marginatum,
Burm. Handb. Ent. ii. 258—Dictyonota pilicornis, *H.-Sch. Wans. Ins.*
iv. 74, pl. 129, f. 401—Tingis (Derephysia) pilicornis, *Kolen. Mel.
Ent.* vi. 14.

a—f. England.
g. Europe. Presented by W. W. Saunders, Esq.

5. DICTYONOTA FIEBERI.

Fieberi, *Förster, Fieb. Eur. Hem.* 127.
Prussia.

6. DICTYONOTA ALBIPENNIS.

albipennis, *Bären. Berl. Ent. Zeit.* 1858, 207, f. 12. *Fieb. Eur. Hem.* 127.

7. DICTYONOTA STRICHNOCERA.

Tingis Eryngii, *Curt. Brit. Ent.* iv. pl. 154 — Dictyonota crassicornis,
H.-Sch. Wanz. Ins. iv. 74, pl. 129, f. B—Dictyonota strichnocera,
Fieb. Ent. Monogr. 95, pl. 8, f. 4—7; *Eur. Hem.* 127. *Dougl. and
Scott, Hem.* 256.

Europe.

8. DICTYONOTA AUBEI.

Dictyonota Aubei, *Sgnt. A. S. E. F. 4me Sér.* v. 118.
South France.

9. DICTYONOTA MARMOREA.

marmorea, *Bärensp. Berl. Ent. Zeit.* 1858, 206, f. 11. *Fieb. Eur. Hem.*
127.
Andalusia.

10. DICTYONOTA PULCHELLA.

pulchella, *Costa Additamenta ad Centurias Cimicum Regni Neapolitani,*
    9; *Atti. Accad. Scienze Napoli*, i. pl. 3, f. 7.
Calabria.

11. DICTYONOTA ERYTHROCEPHALA.

erythrocephala, *Garbiglietti, Bull. Soc. Ent. Ital.* i. 275.
North Italy.

Div. 3. DERYPHISIA.

Deryphisia, *Spin. Ess. Gen.* 73. *Fieb. Ent. Mon.* 99, Gen. 12, pl. 8, f.
    23, 33; *Eur. Hem.* 128. *Dougl. and Scott, Hem.* 253.

12. DICTYONOTA FOLIACEA.

Tingis foliacea, *Fall. Hem. Suec.* 149. *H.-Sch. Panz. Faun. Germ.* 118,
    18. *H.-Sch. Wanz. Ins.* iv. 70, pl. 129, f. D., pl. 130, f. M. N.—
    Deryphisia foliacea, *Spin. Ess. Hem.* 106. *Fieb. Ent. Monogr.* 99,
    pl. 8, f. 23, 27; *Eur. Hem.* 128. *Dougl. and Scott, Hem.* 254.

a, b. England.   Presented by F. Walker, Esq.
c, d. England.

13. DICTYONOTA CRISTATA.

Tingis cristata, *Panz. Faun. Germ.* 99, 19. *H.-Sch. Wanz. Ins.* iv. pl. 130,
    f. 1, L. *Burm. Handb. Ent.* ii. 259—Deryphisia cristata, *Fieb. Ent.
    Monogr.* 100, pl. 8, f. 28, 33 ; *Eur. Hem.* 128.

a—e. England.

South Asia.

14. DICTYONOTA CINGALENSIS.

*Flavescente testacea ; caput bicarinatum ; antennæ nigricantes, crassæ,
    corporis dimidio longiores ; prothorax tricarinatus, antice angustus,
    postice productus et acutus ; alæ anticæ reticulatæ, venis ex parte
    nigris.*

Yellowish cinereous.   Head with two keels, which converge from
between the eyes to the hind border. Eyes black. Antennæ blackish,
thick, rather more than half the length of the body ; second joint shorter
than the first ; third more than twice the length of the first and second
together ; fourth fusiform, a little more than half the length of the third.
Prothorax narrowed in front, elongate and acute hindward, with a transverse furrow near the fore border, and with three keels. Legs piceous.
Fore wings minutely reticulated ; the reticulation partly black. Length of
the body 1¾ line.

a—g. Ceylon.   Presented by Dr. Thwaites.

Genus 2. TINGIS.

Tingis, *Fieb. Syst. Rhyn. Serv. Hist. Hem.* 296. *Burm. Handb. Ent.* ii. 259.

Div. 1.

Tingis, *Fieb. Eur. Hem.* 128.

1. TINGIS PYRI.

Acanthia Pyri, *Geoff. Ins.*—Cimex appendiceus, *Vill. Ent.* 488, pl. 3, f. 19—Tingis Pyri, *Fabr. Syst. Rhyn.* 126. *Burm. Handb. Ent.* ii. 259. *H.-Sch. Wanz. Ins.* iv. 68, pl. 126, f. 395. *Blanch. Hist. Nat. Ins.* iii. 112; *Hem.* pl. 2, f. 7. *Serv. Hist. Hem.* 297. *Fieb. Eur. Hem.* 129.

a. England.

2. TINGIS CHLOROPHANA.

chlorophana, *Fieb. Eur. Hem.* 129.

Portugal.

3. TINGIS SPINIFRONS.

spinifrons, *Fall. Cim. Suec.* 148. *Germ. Ahr. Faun. Eur.* 13, 18. *H.-Sch. Wanz. Ins.* iv. 67, pl. 130, f. A—C, G, H. *Serv. Hist. Hem.* 297. *Fieb. Ent. Monogr.* 105, pl. 9, f. 6—12; *Eur. Hem.* 129—Galeatus spinifrons, *Curt.*

a. England.

4. TINGIS AFFINIS.

affinis, *H.-Sch. Wanz. Ins.* iii. 73, pl. 95, f. 290. *Fieb. Ent. Monogr.* 106, pl. 6, f. 13—16; *Eur. Hem.* 129.

Germany.

5. TINGIS SINUATA.

sinuata, *H.-Sch. Wanz. Ins.* iv. 68, pl. 126, f. 394. *Fieb. Ent. Monogr.* 108, pl. 9, f. 22; *Eur. Hem.* 129.

Silesia. Hungary.

6. TINGIS MACULATA.

maculata, *H.-Sch. Wanz. Ins.* iv. 68, pl. 126, f. 393. *Fieb. Ent. Monogr.* pl. 9, f. 21; *Eur. Hem.* 130—Pyri, *H.-Sch. Wanz. Ins.* iii. 74, pl. 95, f. 291—subglobosa, *H.-Sch. Wanz. Ins.* iv. 68. *Fieb. Ent. Monogr.* 106, pl. 9, f. 17—20—Dictyonota Oberti? *Kol. Mel. Ent. Sp.* 216—Tingis cristata? *Cuv. R. An.* 30. *Ins.* 4, pl. 91, f. 5.

Germany.

North America.

7. TINGIS CILIATA.

ciliata, *Say, Works ed. Leconte*, i. 348.
United States.

8. TINGIS ARCUATA.

arcuata, *Say, Works ed. Leconte*, i. 350.
United States.

9. TINGIS HYALINA.

Tingis hyalina, *Mus. Berl. H.-Sch. Wanz.* v. 84, pl. 173, f. 532.
Carolina.

10. TINGIS TILIÆ.

Tiliæ, *Walsh, Proc. Ent. Soc. Phil.* iii. 408.
Illinois.

11. TINGIS AMORPHÆ.

Amorphæ, *Walsh, Proc. Ent. Soc. Phil.* iii. 409.
Illinois.

12. TINGIS GOSSYPII.

Acanthia Gossypii, *Fabr. Ent. Syst.* iv. 78—Tingis Gossypii, *Fabr. Syst. Rhyn.* 259. *Burm. Handb. Ent.* ii. 259. *H.-Sch. Wanz. Ins.* v. 85, pl, 173, f. 534—Galeatus Gossypii, *Stal, Hem. Fabr.* i. 93.
Mexico. St. Thomas.

13. TINGIS FUSCIGERA.

fuscigera, *Stal, Stett. Ent. Zeit.* xxiii. 323.
Mexico.

14. TINGIS DECENS.

decens, *Stal. Stett. Ent. Zeit.* xxiii. 324.
Tabasco, Mexico.

West Indies.

15. TINGIS SIDÆ.

Acanthia Sidæ, *Fabr. Ent. Syst.* iv. 77—Tingis Sidæ, *Fabr. Syst. Rhyn.* 126—Tingis (Tropidocheila) Sidæ, *Stal, Hem. Fabr.* i. 92.
West Indies.

South America.

16. TINGIS ELEVATA.

Aradus elevatus, *Fabr. Syst. Rhyn.* 120—Tingis (Tropidocheila) elevata, *Stal, Hem. Fabr.* i. 91.

South America.

17. TINGIS FUSCOMACULATA.

fuscomaculata, *Stal, Rio Jan. Hem.* 63.

Rio Janeiro.

18. TINGIS SEXNEBULOSA.

sexnebulosa, *Stal, Rio Jan. Hem.* 64.

Rio Janeiro.

19. TINGIS MONACHA.

monacha, *Stal, Rio Jan. Hem.* 64.

Rio Janeiro.

20. TINGIS? INFLATA.

inflata, *Stal, Rio Jan. Hem.* 64.

Rio Janeiro.

21. TINGIS? GLOBIFERA.

globifera, *Stal, Rio Jan. Hem.* 65.

Rio Janeiro.

22. TINGIS MITRATA.

mitrata, *Stal, Rio Jan. Hem.* 64.

Rio Janeiro.

23. TINGIS STEINI.

Steini, *Stal, Rio Jan. Hem.* 64.

Rio Janeiro.

24. TINGIS INDIGENA.

indigena, *Wltn. Ann. Nat. Hist. 3rd Ser.* i. 124.

a—h. Madeira. From Mr. Wollaston's collection.

South Asia.

25. TINGIS ERUSA.

Picea; prothorax crista alta angulata, membranis lateralibus latis angulosis oblique ascendentibus basi subdiaphanis; pedes gracillimi; alæ anticæ latæ, plagis duabus diaphanis, 2a strigam piceam arcuatam includente, costa basi rotundata biangulata.

Piceous. Head slightly elongated. Third joint of the antennæ very slender, about thrice the length of the first and second together. Prothorax with a high angular crest in the disk, and with a large foliaceous obliquely-ascending appendage on each side; this appendage is nearly pellucid towards the base, and has four or five angles along the border. Legs very slender. Fore wings broad, minutely reticulated, with a pellucid patch near the base of the costa, and with a larger patch extending from the exterior part of the costa to the disk, and including a curved piceous streak; costa rounded, and with two obtuse angles near the base. Length of the body 3 lines.

a. Hindostan. Presented by J. C. Bowring, Esq.

26. Tingis alicollis.

Fœm. *Testaceo-cinerea; caput subtus bicarinatum; antennæ testaceæ, setulosæ, corporis dimidio æquilongæ, articulo 4o nigro basi testaceo; prothorax alte unicristatus, lateribus foliaceis oblique ascendentibus valde dilatatis; alæ anticæ latæ, fascia arcuata strigaque lata obliqua fuscis, margine exteriore truncato.*

Female. Testaceous-cinereous. Head with two parallel keels beneath. Antennæ testaceous, setulose, slender, about half the length of the body; first and second joints stout; third nearly thrice the length of the first and second together; fourth black, testaceous towards the base, stouter than the third, and about one-third of its length. Prothorax minutely reticulated, with a high crest, which projects slightly over the head; sides much dilated, leaf-like, obliquely ascending. Legs slender. Fore wings broad, minutely reticulated; costa much rounded near the base; exterior border truncated; a curved brown band near the base, and an oblique brown streak extending from the tip to the disk. Length of the body 2 lines.

a. Hindostan. Presented by W. W. Saunders, Esq.

27. Tingis globulifera.

Cinerea, subtus nigra, antennis pedibusque flavescentibus; prothorax globosus, fusco maculatus, sulco medio valde determinato; alæ anticæ diaphanæ, bituberculatæ, nigro reticulatæ.

Cinereous, black beneath, nearly elliptical. Antennæ and legs pale yellow. Antennæ slender, subclavate, a little longer than the prothorax. Prothorax high and globose on each side; the globes brown, spotted, and parted by a deep furrow. Scutellum elongated. Fore wings pellucid, extending to the tip of the abdomen, regularly reticulated, each with two large tubercles; reticulation black. Length of the body 1¼ line.

a—h. Madras. Presented by Sir W. Elliot. "Lives on the heliotrope."

Country unknown.

28. Tingis cyathicollis.

cyathicollis, *Costa, Ann. Mus. Zool. Nap.* ii. 146, pl. 2, f. 4.

Genus 3. MONANTHIA.

Monanthia, *St. Farg. et Serv. Enc.* x. 653. *Burm. Handb. Ent.* ii. 260.
Serv. Hist. Hem. 298. *Dougl. and Scott, Hem.* 243.

Europe.
Div. 1.

Dictyonota, *Fieb. Eur. Hem.* 119.

Subdiv. 1.

Platychilæ, *Fieb.*

1. MONANTHIA GRISEA.

Tingis grisea, *Germ. Faun. Eur.* 15, 13—Monanthia grisea, *H.-Sch. Wanz.
Ins.* iv. 60, pl. 125, D. *Fieb. Ent. Monogr.* 64, pl. 5, f. 25—27; *Eur.
Hem.* 120—Deryphisia crispata, *H.-Sch. Wanz. Ins.* iv. 72, pl. 128,
f. 399. *Fieb. Ent. Monogr.* 66, pl. 5, f. 28—30.

a, b. England.

2. MONANTHIA SINUATA.

sinuata, *Fieb. Ent. Monogr.* 60, pl. 5, f. 12—15; *Eur. Hem.* 120—Cardui,
H.-Sch. Wanz. Ins. iv. 61, pl. 127, f. B — Cataplatus auriculatus,
Costa, Cent. 1848, 255.

Europe.

3. MONANTHIA AMPLIATA.

ampliata, *Fieb. H.-Sch. Wanz. Ins.* iv. 62, pl. 127, f. 397 a. *Fieb. Ent.
Monogr.* 59, pl. 5, f. 10, 11; *Eur. Hem.* 120. *Dougl. and Scott,
Hem.* 252.

a. England.

4. MONANTHIA CARDUI.

Cimex Cardui, *Linn. Faun. Suec.* 920; *Syst. Nat.* ii. 718. *Deg. Ins.* iii.
309, pl. 16, f. 1—6—Acanthia Cardui, *Fabr. Ent. Syst.* iv. 77. *Wolff.
Icon. Cim.* 45, pl. 5, f. 42. *Fanz. Faun. Germ.* 3, 24—Tingis Cardui,
Fabr. Syst. Rhyn. 125. *Fall. Hem. Suec.* i. 143. *Zett. Ins. Lapp.*
269—Acanthia clavicornis, *Panz. Faun. Germ.* 3, 24—Monanthia
Cardui, *Burm. Handb. Ent.* ii. 260. *Serv. Hist. Hem.* 298. *H.-Sch.
Wanz. Ins.* iv. 61, pl. 127, f. A, B. *Fieb. Ent. Monogr.* 61, pl. 5,
f. 1—8; *Eur. Hem* 120. *Sahlb. Geoc. Fen.* 131. *Dougl. and Scott,
Hem.* 251—Monanthia (Phyllontocheila) Cardui, *Flor. Rhyn. Liv.*
i. 345.

a. England. Presented by J. C. Dale, Esq.
b—k. England.

5. MONANTHIA COGNATA.

cognata, *Fieb. Eur. Hem.* 121.
Corsica.

6. MONANTHIA ANGUSTATA.

angustata, *H.-Sch. Wanz. Ins.* iv. 61, pl. 127, f. 397 b.   *Fieb. Ent. Monogr.*
62, pl. 5, f. 16—18; *Eur. Hem.* 121.

Germany.

7. MONANTHIA RAGUSANA.

Ragusana, *Kust. Fieb. Eur. Hem.* 121.

Dalmatia.

8. MONANTHIA ELONGATA.

elongata, *Fieb. Eur. Hem.* 121.

Servia.

9. MONANTHIA BRACHYCERA.

brachycera, *Fieb. Eur. Hem.* 121.

Servia.

10. MONANTHIA ECHINOPSIDIS.

Tingis testacea, *H.-Sch. Panz. Faun. Germ.* 118, 23—Monanthia testacea,
H-Sch. Wanz. Ins. iv. 60, pl. 125, f. H. i.—Monanthia Echinopsidis,
Fieb. Ent. Monogr. 62, pl. 5, f. 19—22.

Europe.

11. MONANTHIA SETULOSA.

Tingis capucina, *Germ. Ahr. Faun. Eur.* 18, 24—Tingis gracilis, *H.-Sch.
Panz. Faun. Germ.* 18, 21—Deryphisia gracilis, *H.-Sch. Wanz. Ins.*
iv. 72—Monanthia setulosa *var.* a. capucina, *Fieb. Ent. Monogr.* 58,
pl. 5, f. 34—36—Monanthia setulosa *var.* b. gracilis, *Fieb. Ent.
Monogr.* 69, pl. 5, f. 34—38; *Eur. Hem.* 122.

Europe.

12. MONANTHIA CILIATA.

Tingis reticulata, *H.-Sch. Wanz. Ins.* iii. 72, pl. 95, f. 288—Deryphysia
reticulata, *H.-Sch. Wanz. Ins.* iv. 71—Monanthia ciliata, *Fieb. Ent.
Monogr.* 67, pl. 5, f. 31—33; *Eur. Hem.* 122—Tingis ciliata, *Spin.
Ess. Hem.* 166—Monanthia reticulata, *Dougl. and Scott, Hem.* 250.

a. Europe. Presented by W. W. Saunders, Esq.
b. England?

Subdiv. 2. Tropidochila, *Fieb.*

13. MONANTHIA PILOSA.

Monanthia angusticollis, *H.-Sch. Wanz. Ins.* iii. 72, pl. 95, f. 289—
Monanthia pilosa, *Fieb. Ent. Monogr.* 79, pl. 6, f. 36, 37; *Eur. Hem.*
122—Monanthia villosa, *Costa, Cent.* pl. 6, f. 6—Deryphysia reticulata,
Spin. Ess. Hem. 166.

a. Europe. Presented by W. W. Saunders, Esq.

14. MONANTHIA COSTATA.

Acanthia costata, *Fabr. Ent. Syst.* iv. 77—Tingis costata, *Fabr. Syst. Rhyn.* 152. *Fall. Hem. Suec.* 143. *Germ. Faun. Eur.* 18, 25—Monanthia costata, *Burm. Handb. Ent.* ii. 261. *H.-Sch. Wanz. Ins.* iv. 55, pl. 123, f. 390. *Fieb. Ent. Monogr.* 72, pl. 6, f. 10—12; *Eur. Hem.* 123. *Sahlb. Geoc. Fen.* 132. *Dougl. and Scott, Hem.* 249—Laccometopus costatus, *Stal, Hem. Fabr.* i. 92—Monanthia (Tropidocheila) costata, *Flor, Rhyn. Liv.* i. 347.

a—h. England.

15. MONANTHIA CRASSIPES.

crassipes, *Fieb. Eur. Hem.* 123.

Servia.

16. MONANTHIA LITURATA.

liturata, *Fieb. Ent. Monogr.* 74, pl. 6, f. 16—18; *Eur. Hem.* 123.

Spain.

17. MONANTHIA STACHYDIS.

Tingis grisea, *H.-Sch. Nom.* 58—Monanthia Stachydis, *Fieb. Ent. Monogr.* 73, pl. 6, f. 13—15; *Eur. Hem.* 123—Monanthia maculata, *H.-Sch. Wanz. Ins.* iv. 56, pl. 123, f. 389.

Europe.

18. MONANTHIA GENICULATA.

geniculata, *Fieb. Ent. Monogr.* 75, pl. 6, f. 19—21; *Eur. Hem.* 124.

Austria. Hungary. Carinthia.

Subdiv. 3. Physatochilæ, *Fieb.*

19. MONANTHIA ALIENA.

aliena, *Fieb. Eur. Hem.* 124.

Turkey. Syria.

20. MONANTHIA ERYNGII.

Tingis Eryngii, *Latr. Hist. Ins.* xiii. 253—Tingis melanocephala, *Panz. Faun. Germ.* 100, 21—Piesma melanocephala, *Burm. Handb. Ent.* ii. 258—Monanthia melanocephala, *Fieb. Ent. Monogr.* 77, pl. 6, f. 26—30; *Eur. Hem.* 124.

Enrope.

21. MONANTHIA ALBIDA.

albida, *H.-Sch. Wanz. Ins.* iv. 54, pl. 126, f. 306, pl. 125 P. *Fieb. Eur. Hem.* 124—Schæfferi, *Fieb. Ent. Monogr.* 78, pl. 6, f. 31—35.

East Germany.

22. Monanthia quadrimaculata.

Acanthia quadrimaculata, *Wolff. Icon. Cim.* 132, pl. 13, f. 127—Tingis
quadrimaculata, *Fall. Hem. Suec.* 144—Tingis corticea, *H.-Sch.*
Fanz. Faun. Germ. 118, 22—Monanthia quadrimaculata, *Burm.*
Handb. Ent. ii. 261. *H.-Sch. Wanz. Ins.* iv. 58, pl. 125, f. A.
Fieb. Ent. Monogr. 81, pl. 7, f. 1—3; *Eur. Hem.* 124. *Dougl. and*
Scott, Hem. 247—Monanthia dumetorum, *Sahlb. Geoc.* 132—Monan-
thia (Physatocheila) quadrimaculata, *Flor, Rhyn. Liv.* i. 550.

Europe.

23. Monanthia dumetorum.

dumetorum, *H.-Sch. Wanz. Ins.* iv. 57, pl. 124, f. 391—*Fieb. Ent.*
Monogr. 82, pl. 27, f. 4—6; *Eur. Hem.* 125. *Dougl. and Scott, Hem.*
246—Tingis Oxyacanthæ? *Curt. Brit. Ent.* xvi. pl. 741.

Europe.

24. Monanthia scapularis.

Tingis simplex, *Panz. Faun. Germ.* 118, 21—Monanthia simplex, *H.-Sch.*
Wanz. Ins. iv. 59, pl. 128, f. F. *Dougl. and Scott, Hem.* 245—
Monanthia scapularis, *Fieb. Ent. Monogr.* 80, pl. 6, f. 38—10 ; *Eur.*
Hem. 125.

Europe.

25. Monanthia platyoma.

platyoma, *Fieb. Eur. Hem.* 125.

Austria.   Bohemia.

26. Monanthia Wolffi.

Acanthia Echii, *Wolff. Icon. Cim.* 130, pl. 13, f. 124—Tingis Humuli,
Fall. Hem. Suec. 144—Monanthia Humuli, *Burm. Handb. Ent.* ii.
261—Monanthia Echii, *H.-Sch. Wanz. Ins.* iv. 14, pl. 114, f. 360—
Monanthia Wolffi, *Fieb. Ent. Monogr.* 86, pl. 22—24 ; *Eur. Hem.*
125.

Europe.

27. Monanthia Humuli.

Acanthia Humuli, *Fabr. Ent. Syst.* iv. 77—Tingis Humuli, *Fabr. Syst.*
Rhyn. 126—Monanthia convergens, *Klug. Burm. Handb. Ent.* ii. 261.
H.-Sch. Wanz. Ins. iv. 15, 58, pl. 114, f. 361—Monanthia Humuli,
Fieb. Ent. Monogr. 84, pl. 7, f. 17, 18; *Eur. Hem.* 125. *Dougl. and*
Scott, Hem. 244—Monanthia (Physatocheila) Humuli, *Flor, Rhyn.*
Liv. i. 335.

a. England.   Presented by F. Walker, Esq.
b, c. England.

28. Monanthia Lupuli.

Lupuli, *Kunze, Fieb. Ent. Monogr.* 85, pl. 7, f. 19, 20 ; *Eur. Hem.* 126.
H.-Sch. Wanz. Ins. iv. 13, pl. 114, f. 359.

Germany.

29. MONANTHIA VESICULIFERA.

costata, *H.-Sch. Wanz. Ins.* iv. 15, pl. 114, f. 362. *Burm. Hand. Ent.* ii. 261—vesiculifera, *Fieb. Ent. Monogr.* 87, pl. 7, f. 25, 26 ; *Eur. Hem.* 126.
Europe.

30. MONANTHIA ECHII.

Tingis Echii, *Fabr. Syst. Rhyn.* 126—Monanthia rotundata, *H.-Sch. Wanz. Ins.* iv. 58, pl. 120 É ; pl. 124, f. 392—Monanthia Echii, *Fieb. Ent. Mon.* 88, pl. 7, f. 27—32.
Europe.

Subdiv. ?
31. MONANTHIA UNICOSTATA.

unicostata, *Muls. Ann. Soc. Linn.* 1852, 134.
France.

32. MONANTHIA KIESENWETTERI.

Kiesenwetteri, *Muls. Ann. Soc. Linn.* 1852, 135.
France.

33. MONANTHIA RETICULATA.

reticulata, *Ramb. Faun. And.*
Spain.

34. MONANTHIA PARELLELA.

Monanthia (Cataplatus) parallela, *Costa, Cim. R. Neap. Cent.*
South Italy.

35. MONANTHIA VARIOLOSA.

Monanthia (Cataplatus) variolosa, *Costa, Cim. R. Neap. Cent.*
South Italy.

36. MONANTHIA SIMILIS.

similis, *Dougl. and Scott, Ent. M. Mag.* v. 259.
England.

37. MONANTHIA PALLIDA.

pallida, *Garbiglietti, Bull. Soc. Ent. Ital.* i. 273.
North Italy.

38. MONANTHIA PILIGERA.

piligera, *Gar. Bull. Soc. Ent. Ital.* i. 273.
North Italy.

39. MONANTHIA LURIDA.

lurida, *Gar. Bull. Soc. Ent. Ital.* i. 274.
North Italy.

40. MONANTHIA OBLONGA.

oblonga, *Gar. Bull. Soc. Ent. Ital.* i. 274.
North Italy.

41. MONANTHIA UNICOLOR.

unicolor, *Gar. Bull. Soc. Ent. Ital.* i. 274.
Sardinia.

Div. 2.

Laccometopus, *Fieb. Ent. Monogr.* 96 ; *Eur. Hem.* 119.

42. MONANTHIA CLAVICORNIS.

Cimex clavicornis, *Linn. Faun. Suec.* 911—Tingis clavicornis, *Fabr. Syst. Rhyn.* 124. *Panz. Faun. Germ.* 23, 23—Laccometopus clavicornis, *Fieb. Ent. Monogr.* 97, pl. 8, f. 10—16 ; *Eur. Hem.* 119. *Frauenf, Zool. Bot. Ver.* 1853, 157.
a, b. England.

43. MONANTHIA TEUCRII.

Cimex Teucrii, *Host, Coll.* ii. 255, pl. 18—Laccometopus Teucrii, *Frauenf. Zool. Bot. Verh.* 1853, 157. *Fieb. Eur. Hem.* 119.
Austria. Italy.

Div. 3.

Orthosteira, *Fieb. Ent. Monogr.* 46 ; *Eur. Hem.* 130. *Dougl. and Scott, Hem.* 260.

44. MONANTHIA CASSIDEA.

Tingis cassidea, *Fall. Hem. Suec.* 146—Tingis brunnea, *Germ. Faun. Eur.* 18, 23. *H.-Sch. Wanz. Ins.* iv. 25, pl. 118, f. 374—Orthosteira cassidea, *Fieb. Ent. Monogr.* 47, pl. 3, f. 39—42—Orthosteira brunnea, *Fieb. Ent. Monogs.* pl. 3, f. 43, 44—Orthostira cassidea, *Fieb. Eur. Hem.* 130.
a—f. England.

45. MONANTHIA CERVINA.

Tingis cervina, *Germ. Faun. Ins. Eur.* 18, 22—Monanthia cervina, *H.-Sch. Wanz. Ins.* iv. 26, 53, pl. 118, f. 375, pl. 129, f. G—Orthosteira cervina, *Fieb. Ent. Monogr.* 48, pl. 4, f. 1—3. *Suhlb. Geoc. Fen.* 129—Orthostira cervina, *Fieb. Eur. Hem.* 130. *Dougl. and Scott, Hem.* 262—Monanthia (Orthosteira) cervina, *Flor, Rhyn. Liv.* i. 341—Orthosteira platycheila, *Fieb. Ent. Monogr.* 53, pl. 4, f. 15—18 — Orthostira platycheila, *Fieb. Eur. Hem.* 130.

Europe.

46. MONANTHIA GRACILIS.

Orthosteira gracilis, *Fieb. Ent. Monogr.* 54, pl. 4, f. 19—21—Orthostira gracilis, *Fieb. Eur. Hem.* 131.

Germany.

47. MONANTHIA OBSCURA.

obscura, *H.-Sch. Wanz. Ins.* iv. 23, pl. 118, f. 372—pusilla, *Burm. Handb. Ent.* ii. 262—Orthosteira obscura, *Fieb. Ent. Monogr.* 54, pl. 4, f. 22—25. *Sahlb. Geoc. Fen.* 130—Orthostira obscura, *Fieb. Eur. Hem.* 131. *Dougl. and Scott, Hem.* 263.

Europe.

48. MONANTHIA NIGRINA.

Tingis nigrina, *Fall. Hem. Suec.* 145. *Panz. Faun. Germ.* 118, 16—Monanthia nigrina, *Fieb. Ent. Monogr.* pl. 5, f. 23, 24. *Eur. Hem.* 131—Orthosteira cinerea, *Fieb. Ent. Monogr.* 52, pl. 4, f. 11—14. *Sahlb. Geoc. Fen.* 130.

a—l. England.

49. MONANTHIA PUSILLA.

Acanthia marginata, *Wolff, Icon. Cim.* f. 126—Tingis carinata, *Panz. Faun. Germ.* 99, 20—Tingis pusilla, *Fall. Hem. Suec.* 146. *Wanz. Ins.* iv. 24, pl. 118, f. 373—Orthosteira pusilla, *Fieb. Ent. Monogr.* 51, pl. 4, f. 9—Orthosteira macrophthalma, *Fieb. Ent. Monogr.* 49, pl. 4, f. 4—7—Orthostira pusilla, *Fieb. Eur. Hem.* 131.

a—f. England.

50. MONANTHIA CONCINNA.

Orthostira concinna, *Fieb. Wien. Ent. Mon.* viii. 211. *Dougl. and Scott, Hem.* 260.

England.

Div. 4.

Campylosteira, *Fieb. Ent. Monogr.* 42—Campylostira, *Fieb. Eur. Hem.* 131. *Dougl. and Scott, Hem.* 257.

51. MONANTHIA CILIATA.

Campylosteira ciliata, *Fieb. Ent. Monogr.* 93, pl. 3, f. 27—32—Campylostira ciliata, *Fieb. Eur. Hem.* 132.

Bohemia.

52. MONANTHIA FALLENI.

Campylosteira Falleni, *Fieb. Ent. Monogr.* 43, pl. 3, f. 23—26—Campylostira Falleni, *Fieb. Eur. Hem.* 132.

Germany.

53. MONANTHIA BRACHYCERA.

Orthosteira brachycera, *Fieb. Ent. Monogr.* 43, pl. 3, f. 27—33. *Fieb.
Eur. Hem.* 132. *Dougl. and Scott, Hem.* 259.

Europe.

54. MONANTHIA SINUATA.

Campylostira sinuata, *Fieb. Eur. Hem.* 132.

Germany.

55. MONANTHIA VERNA.

Tingis verna, *Fall. Hem. Suec.* 147—Monanthia verna, *H.-Sch. Wanz.
Ins.* iv. 64, pl. 127, f. 398—Campylosteira verna, *Fieb. Ent. Monogr.*
45, pl. 3, f. 38—Campylostira verna, *Fieb. Eur. Hem.* 132. *Dougl.
and Scott, Hem.* 258.

Europe.

56. MONANTHIA PARVULA.

Monanthia (Monosteira) parvula, *Sgnt. A. S. E. F. 4me Sér.* v. 117.

South France.

North America.

57. MONANTHIA CINEREA.

Tingis cinerea, *Say, Works ed Leconte,* i. 349.

United States.

58. MONANTHIA MUTICA.

Tingis mutica, *Say, Works ed Leconte,* i. 349.

United States.

59. MONANTHIA PLEXA.

Tingis plexus, *Say, Works ed Leconte,* i. 349.

United States.

Mexico.

60. MONANTHIA SACCHARI.

Acanthia Sacchari, *Fabr. Ent. Syst.* iv. 77—Tingis Sacchari, *Fabr. Syst. Rhyn.* 126—Monanthia Sacchari, *H.-Sch. Wanz. Ins.* v 85, pl. 73, f. 533. *Fieb. Ent. Monogr.* 76, pl. 6, f. 22—25—Tingis (Tropidocheila) Sacchari, *Stal, Hem. Fabr.* i. 92.

Mexico. West Indies. Brazil.

61. MONANTHIA TABIDA.

Monanthia tabida, *Mus. Berol. H.-Sch. Wanz.* v. 86, pl. 173, f. 535.
Mexico.

62. MONANTHIA PATRICIA.

M. (Phyllontocheila) patricia, *Stal, Stett. Ent. Zeit.* xxiii. 324—M. (Gargaphia) patricia, *Stal, Stett. Ent. Zeit.* xxiii. 324.
Mexico.

63. MONANTHIA LUCIDA.

Testacea, gracilis, glabra, sublinearis ; prothorax antice et caput nigricantia ; alæ anticæ diaphanæ, fasciis tribus nigricantibus, 1a obliqua, 2a 3aque arcuatis.

Testaceous, slender, smooth, shining, nearly linear. Head small, blackish. Eyes rather prominent. Prothorax blackish in front. Fore wings pellucid, with an oblique blackish band near the base and with two curved blackish bands beyond the middle, the outer one very near the exterior border ; reticulation large ; costa forming a prominent rounded angle near the base. Length of the body 1¼ line.

The specimen described is mutilated.

a. Vera Cruz. Presented by W. W. Saunders, Esq.

West Indies.

64. MONANTHIA SIDÆ.

Acanthia Sidæ, *Fabr. Ent. Syst.* iv. 77— Tingis Sidæ, *Fabr. Syst. Rhyn.* 126—Tingis (Tropidocheila) Sidæ, *Stal, Hem. Fabr.* i. 92.
West Indies.

South America.

Div. Tropidocheila.

65. MONANTHIA ELEVATA.

Aradus elevatus, *Fabr. Syst. Rhyn.* 120—Tingis (Tropidocheila) elevata, *Stal, Hem. Fabr.* i. 91.
South America.

Div. Gargaphia.

66. MONANTHIA TRICOLOR.

M. (Gargaphia) tricolor, *Mayr, Verh. Zool. Bot. Gesell. Wien.* xii. 442.
Venezuela.

67. MONANTHIA MUNDA.

munda, *Stal, Rio Jan. Hem.* 60—M. (Gargaphia) munda, *Stal, Stett. Ent.*
Zeit. xxiii. 324.
Rio Janeiro.

68. MONANTHIA FORMOSA.

formosa, *Stal, Rio Jan. Hem.* 61—M. (Gargaphia) formosa, *Stal, Stett.*
Ent. Zeit. xxiii. 324.
Rio Janeiro.

69. MONANTHIA FLEXUOSA.

flexuosa, *Stal, Rio Jan. Hem.* 61—M. (Gargaphia) flexuosa, *Stal, Stett.*
Ent. Zeit. xxiii. 324.
Rio Janeiro.

70. MONANTHIA SIMULANS.

simulans, *Stal, Rio Jan. Hem.* 61—M. (Gargaphia) simulans, *Stal, Stett.*
Ent. Zeit. xxiii. 324.
Rio Janeiro.

71. MONANTHIA ARMIGERA.

armigera, *Stal, Rio Jan. Hem.* 61.
Rio Janeiro.

72. MONANTHIA SPINULIGERA.

spinuligera, *Stal, Rio Jan. Hem.* 61.
Rio Janeiro.

73. MONANTHIA PALLIPES.

pallipes, *Stal, Rio Jan. Hem.* 62.
Rio Janeiro.

74. MONANTHIA MARGINELLA.

marginella, *Stal, Rio Jan. Hem.* 62.
Rio Janeiro.

75. MONANTHIA OCHROPA.

ochropa, *Stal, Rio Jan. Hem.* 62.
Rio Janeiro.

76. MONANTHIA FUSCOCINCTA.

fuscocincta, *Stal, Rio Jan. Hem.* 62.
Rio Janeiro.

77. MONANTHIA DOHRNI.

Dohrni, *Stal, Rio Jan. Hem.* 62.
Rio Janeiro.

78. MONANTHIA APPROXIMATA.

approximata, *Stal, Rio Jan. Hem.* 63.
Rio Janeiro.

79. MONANTHIA LEPIDA.

lepida, *Stal, Rio Jan. Hem.* 63.
Rio Janeiro.

80. MONANTHIA MONOTROPIDIA.

monotropidia, *Stal, Rio Jan. Hem.* 63.
Rio Janeiro.

Div. Laccometopus.
81. MONANTHIA ALBILATERA.

Laccometopus albilaterus, *Stal, Rio Jan. Hem.* 65.
Rio Janeiro.

82. MONANTHIA PROLIXA.

Laccometopus prolixus, *Stal, Rio Jan. Hem.* 65.
Rio Janeiro.

83. MONANTHIA MORIO.

Laccometopus morio, *Stal, Rio Jan. Hem.* 65.
Rio Janeiro.

84. MONANTHIA LUCTUOSA.

Laccometopus luctuosus, *Stal, Rio Jan. Hem.* 65.
Rio Janeiro.

85. MONANTHIA LUNULATA.

lunulata, *Mayr, Verh. Zool. Bot. Gesell. Wien.* xv. 441. *Reise Novara,*
   Hem. 163, f. 46.
Rio Janeiro.

Div. Solenostoma.

Solenostoma, *Sgnt. A. S. E. F.* 4me *Sér.* iii. 575.

86. MONANTHIA LILIPUTIANA.

Solenostoma Liliputiana, *Sgnt. A. S. E. F.* 4me *Sér.* iii. 575, pl. 13, f. 27.
Chili.

Div. n.

87. MONANTHIA LANCEOLATA.

*Nigra, longi-fusiformis ; prothorax tricarinatus, postice attenuatus,
margine antico testaceo, carina media alta postice humuli ; pedes
longiusculi ; alæ anticæ striga costali albida diaphana.*

Black, slender, elongate-fusiform. Head slightly elongate. Prothorax
narrower in front, with a transverse furrow, with a testaceous fore border,
and with three keels; middle keel very high, testaceous in front, testaceous
and much lower in the hind part, which is elongated and acute. Legs
slender, rather long. Fore wings opaque, minutely reticulated, with a
whitish diaphanous costal streak, which is attenuated towards the base and
does not extend to the tip. Length of the body 2½ lines.

a. Brazil. Presented by W. W. Saunders, Esq.

88. MONANTHIA LINEIFERA.

*Nigra, fusiformis ; prothorax tricarinatus, lateribus et carina media alta
antice producta albidis diaphanis reticulatis, margine postico
producto acuto apicem versus albido ; pedes longi, graciles, sordide
albidi, femoribus nigricantibus ; alæ anticæ albæ, diaphanæ, lineis
tribus transversis strigaque obliqua arcuata fuscata nigris.*

Black, fusiform. Head slightly elongated. Prothorax with a whitish
diaphanous and reticulated border on each side and with three whitish
keels; middle keel diaphanous, reticulated and very high in front, where it
extends a little over the head; hind part elongated, acute, whitish towards
the tip. Legs long, slender, dingy whitish; femora mostly blackish. Fore
wings white, diaphanous, minutely reticulated, with three transverse black
lines near the base and with an oblique forked curved black streak, which
extends from near the end of the interior border to near the end of the
costa. Length of the body 2 lines.

a. Brazil. Presented by W. W. Saunders, Esq.

Africa.

A. Sides of the prothorax not reflexed. Subg. Phyllontochila, *Fieb. Ent.
Mon.* 59. *Stal, Hem. Afr.* iii. 27.
B. Sides of the prothorax reflexed.
a. Sides of the prothorax very narrowly foliaceous. Subg. Physatochila,
Fieb. Ent. Mon. 80. *Stal, Hem Afr.* iii. 28.
b. Sides of the prothorax very broadly foliaceous. Subg. Elasmognathus,
Fieb. Ent. Mon. 90. *Stal, Hem. Afr.* iii. 29.

89. MONANTHIA NIGRICEPS.

Monanthia nigriceps, *Sgnt. A. S. E. F. 3me Sér.* viii. 95—Monanthia
(Physatochila) nigriceps, *Stal, Hem. Afr.* iii. 29.
Madagascar.

90. MONANTHIA ALATICOLLIS.

Phyllontocheila alaticollis, *Stal, Ofv. Vet. Ak. Forh.* 1855, 37—Monanthia
(Phyllontochila) alaticollis, *Stal, Hem. Afr.* iii. 27.
Caffraria.

91. MONANTHIA SORDIDA.

Monanthia (Physatocheila) sordida, *Stal, Eug. Resa, Hem.* 259—Monanthia
(Physatochila) sordida, *Stal, Hem. Afr.* iii. 29.
Cape.

92. MONANTHIA ORNATELLA.

Tropidocheila ornatella, *Stal, Ofv. Vet. Ak. Forh.* 1855, 37—Monanthia
(Physatochila) ornatella, *Stal, Hem. Afr.* iii. 28.
Caffraria.

93. MONANTHIA FIEBERI.

Elasmognathus Ficheri, *Stal, Ofv. Vet. Ak. Forh.* 1855, 38—Monanthia
(Elasmognathus) Fieberi, *Stal, Hem. Afr.* iii. 29.
Caffraria.

94. MONANTHIA NATALENSIS.

Physatocheila Natalensis, *Stal, Ofv. Vet. Ak. Forh.* 1855, 38—Monanthia
(Physatochila) Natalensis, *Fieb. Hem. Afr.* iii. 28.
Caffraria.

95. MONANTHIA WAHLBERGI.

Phyllontocheila Wahlbergi, *Stal, Ofv. Vet. Ak. Forh.* 1855—Monanthia
(Phyllontochila) Wahlbergi, *Stal, Hem. Afr.* iii. 27.
Caffraria.

South Asia
Div. Elasmognathus.
Elasmognathus, *Fieb. Ent. Monogr. H.-Sch. Wanz. Ins.* ix. 147, 156.

96. MONANTHIA HELFERI.

Helferi, *Fieb. Ent. Monogr.* pl. 7, f. 33—41. *H.-Sch. Wanz. Ins.* ix. 156,
pl. 311, f. 955.
Hindostan.

Australasia.
Div.
Diconocoris, *Mayr, Verh. Zool. Bot. Ver. Wien.* xv. 442.

97. Monanthia Javana.
Diconocoris Javanus, *Mayr, Verh. Zool. Bot. Ver. Wien.* xv. 442.
Java.

98. Monanthia invaria.

Nigra, longi-fusiformis; antennæ graciles, filiformes, corporis dimidio
*   paullo longiores; prothorax subtiliter scaber, subcarinatus, postice*
*   productus reticulatus acutus; pedes longiusculi; alæ anticæ cinereæ,*
*   fusco reticulatæ et bifasciatæ.*

Black, slender, elongate-fusiform. Head slightly elongated. Antennæ
slender, filiform, a little more than half the length of the body; first joint
stout; second much shorter than the first; third more than thrice the
length of the first and second together; fourth much more than half the
length of the third. Prothorax finely scabrous, with a slight keel and with
a distinct transverse furrow near the fore border; hind part elongated,
acute, reticulated. Legs slender, rather long. Fore wings cinereous,
minutely reticulated; reticulation brown; two irregular brown bands near
the base. Length of the body 2½ lines.

a. New Guinea. Presented by W. W. Saunders, Esq.

99. Monanthia monticollis.

Fulva; antennæ gracillimæ, corporis dimidio longiores, articulo 4o atro
*   piloso longi-fusiformi; prothorax nodis duobus magnis piceis reticu-*
*   latis; abdomen piceum; alæ anticæ cinereæ, nigricante reticulatæ et*
*   bimaculatæ.*

Tawny. Head small. Eyes black, small, close to the fore border of
the prothorax. Antennæ very slender, rather more than half the length of
the body; first joint stout; second shorter than the first; third more than
four times the length of the first and second together; fourth deep black,
hairy, elongate-fusiform, stouter than the third and about thrice its length.
Prothorax forming two large globose piceous protuberances, which are
largely reticulated. Abdomen piceous. Legs slender. Fore wings
cinereous, minutely reticulated; reticulations blackish, increasing in size
from the base to the tip; two blackish spots on the costa, one before the
middle, the other near the tip. Length of the body 1¾ line.

a. Sarawak. Presented by W. W. Saunders, Esq.

100. Monanthia irregularis.
Physatocheila irregularis, *Mrtz. A. S. E. F.* 4me Sér. i. 68.
Lifu.

Australia.

101. MONANTHIA VESICULATA.

Monanthia (Physatocheila) vesiculata, *Stal, Eug. Resa*, 259.

Sydney.

102. MONANTHIA GIBBIFERA.

Fæm. *Cinerea, subtus fulva ; antennæ fulvæ, corporis dimidio æqui-longæ, articulo 4o nigro fusiformi ; prothorax tricarinatus, carina media antice altissima; pedes fulvi, graciles ; alæ anticæ nigro reticulatæ.*

Female. Cinereous, tawny beneath, rather broad. Head small. Antennæ tawny, very slender, about half the length of the body; first and second joints short; third very long; fourth black, fusiform, less than one-fourth of the length of the third. Prothorax with a very high ridge on the fore part; this ridge projects a little above the head, and there is a depression between it and the lower ridge, which is on the hind part; a keel on each side. Legs tawny, slender. Fore wings minutely reticulated; reticulation black. Length of the body 1½ line.

a. Australia. Presented by F. Walker, Esq.

Country unknown.

Div. Teleonemia, *Costa, Ann. Mus. Zool. Nap.* ii. 144.

103. MONANTHIA FUNEREA.

Teleonemia funerea, *Costa, Ann. Mus. Zool. Nap.* ii. 145, pl. 2, f. 5.

————?

ERRATA.

———o———

In a few instances among the preceding pages different species of one genus are alike as to name; this is owing to the union of recent divisions into larger genera, and the above cases are noticed with the errata as follows:—

Page 47.
For 17. PACHYPETIS *read* 17. PACHYPELTIS.

Page 54.
For Pantilus *read* Pantilius.

Page 56.
For 27. LOPUS PARTILUS *read* 27. LOPUS PARTITUS.

Page 70.
CAPSUS AMŒNUS.
This species is inadvertently inserted again in page 76.

12. CAPSUS FIEBERI. Nomen bis lectum.

Page 72.
21. CAPSUS BIMACULATUS. Nomen bis lectum.

Page 79.
68. CAPSUS CORUSCUS.
This species is accidentally recorded again in page 90.

Page 90.
126. CAPSUS APICALIS. Nomen bis lectum.

Page 98.
Capsus divisus is previously recorded in page 91.

Page 104.

209. Capsus dilatatus. Nomen bis lectum.

Page 105.

216. Capsus bimaculatus. Nomen bis lectum.

Capsus speciosus and C. ocellatus are inserted again in page 108.

Page 117.

For 271. Capsus limbatus *read* 271. Capsus limbifer.

Page 120.

For 278. Capsus stramineus *read* 278. Capsus pallidifer.

Page 123.

288. Capsus pellucidus. Nomen bis lectum.

289. Capsus collaris. Nomen bis lectum.

Page 124.

290. Capsus costalis. Nomen bis lectum.

Page 125.

For 295. Capsus simulans *read* 295. Capsus dissimulans.

For 296. Capsus tristis *read* 296. Capsus lugens.

Page 127.

307. Capsus pellucidus. (bis lectum).

Page 133.

For 28. Leptomerocoris Putori *read* 28. Leptomerocoris Putoni.

Page 135.

42. Leptomerocoris planicornis.

This name may be annulled; the species is recorded as Eurymerocoris Mali in p. 147.

Page 141.

For Pachyxiphus *read* Pachyxyphus.

Page 149.

20. Eurymerocoris nigripes. Nomen bis lectum.

Page 153.
51. EURYMEROCORIS FIEBERI. Nomen bis lectum.

Page 155.
68. EURYMEROCORIS VITTATUS. Nomen bis lectum.

Page 156.
71. EURYMEROCORIS TIBIALIS. Nomen bis lectum.

Page 157.
82. EURYMEROCORIS PRASINUS. Nomen bis lectum.

85. EURYMEROCORIS MEYERI. Nomen bis lectum.

Page 159.
95. EURYMEROCORIS WHITEI. Nomen bis lectum.

INDEX.

E. NEWMAN, PRINTER, DEVONSHIRE STREET, BISHOPSGATE.

CPSIA information can be obtained
at www.ICGtesting.com
Printed in the USA
BVHW090116211118
533638BV00011B/214/P